Springer-Verlag   6900 Heidelberg 1 · Postfach 1780
                   Telefon (06221) 49101 · Telex 04-61723
                   1000 Berlin 33 · Heidelberger Platz 3
                   Telefon (0311) 822001 · Telex 01-83319

Springer-Verlag   New York, NY 10010 · 175, Fifth Avenue
New York Inc.   Telefon 673-2660

32 Fortschritte der chemischen Forschung
Topics in Current Chemistry

Structure and Transformations

of Organic Molecules

Springer-Verlag
Berlin Heidelberg GmbH 1972

ISBN 978-3-540-05936-3   ISBN 978-3-540-37614-9 (eBook)
DOI 10.1007/978-3-540-37614-9

Contents

Quantum Chemistry of
Nonbenzenoid Cyclic Conjugated Hydrocarbons

Prof. Takeshi Nakajima

Department of Chemistry, North Dakota State University, Fargo, North Dakota, USA,
and Department of Chemistry, Faculty of Science, Tohoku University, Sendai, Japan*

Contents

---

* Present address.

I. Introduction

Nonbenzenoid cyclic conjugated hydrocarbons are conveniently classified into two categories: conjugated hydrocarbons composed of odd-membered rings called, in terminology of molecular orbital theory, nonalternant hydrocarbons, and cyclic polyenes currently known as annulenes.

Of the fundamental nonalternant hydrocarbons, only two prototypes were known about fifteen years ago: *azulene* (XI, Fig. 5), the molecular structure of which was determined by Pfau and Plattner[1] and *fulvene* (XIX) synthesized by Thiec and Wiemann[2]. Early in the 1960's many other interesting prototypes have come to be synthesized. Doering[3] succeeded in synthesizing heptafulvene (XX)[4], fulvalene (XXI) and heptafulvalene (XXIII). Prinzbach and Rosswog[5] reported the synthesis of sesquifulvalene (XXII). Preparation of a condensed bicyclic nonalternant hydrocarbon, heptalene (VII), was reported by Dauben and Bertelli[6]. On the other hand, its 5-membered analogue, *pentalene* (I), has remained, up to the present, unvanquished to many attempts made by synthetic chemists. Very recently, de Mayo and his associates[7] have succeeded in synthesizing its closest derivative, 1-methylpentalene. It is added in this connection that dimethyl derivatives of condensed tricyclic nonalternant hydrocarbons composed of 5- and 7-membered rings (XIV and XV), known as Hafner's hydrocarbons, were synthesized by Hafner and Schneider[8] already in 1958.

As early as about twenty years ago, Pullmans and their associates[9] carried out extensive theoretical studies of electronic structures of fundamental nonalternant hydrocarbons, most of which were unknown at that time, using the simple Hückel molecular orbital theory. It is gratifying to note that this precedence of theoretical investigation has stimulated organic chemists to attempts to synthesize these then desk molecules. The simple Hückel theory has predicted substantial π-electron delocalization energies for these molecules, suggesting that they would have in a greater or lesser degree aromatic stability. Experimental information now available, however, indicates that contrary to this expectation, most of the nonalternant hydrocarbons we know possess no aromatic stability like that of the classical aromatic systems.

In the 1930's Hückel[10] proposed, on the basis of molecular-orbital calculations, a theoretical criterion for aromaticity of cyclic polyenes, known as *Hückel's rule*, which states that cyclic polyenes should be aromatic if, and only if, they contain $4n+2$ π-electrons. At that time only two of such cyclic polyenes were known: benzene and cyclopentadienyl anion, each having six π-electrons and satisfying Hückel's rule. Since then, the validity of Hückel's rule had not been challenged until Sondheimer and his group[11),12)] succeeded in synthesizing the higher members of the $[4n+2]$ annulenes about ten years ago. Various experimental facts concerning $C_{30}H_{30}$[12)], the highest member of the $[4n+2]$ annulene known to date, indicate that it is definitely not aromatic. This seems to suggest that the aromaticity of $[4n+2]$ annulenes should decrease with increasing n and disappear above a certain critical ring size, the situation apparently violating Hückel's rule.

The anomalously reduced stabilities of certain nonalternant hydrocarbons and higher members of $[4n+2]$ annulenes arise from their seemingly peculiar geometrical structures in which a strong bond distortion often accompanied by a molecular-symmetry reduction occurs.

For benzenoid hydrocarbons, it has been well recognized that the molecular symmetry for the ground state is always that suggested by the superposition with equal weight of the equivalent Kekulé-type resonance structures, and that the bond orders calculated using valence-bond or molecular-orbital theories, assuming the apparently-full molecular-symmetry, give the theoretical C—C bond lengths which are in good agreement with experimental values.

On the other hand, it was somewhat amazing to discover that the ground states of certain nonbenzenoid hydrocarbons should not adopt the fully-symmetrical nuclear arrangement expected on the basis of the conventional resonance theory, but rather a less symmetrical nuclear configuration in which the nuclei have been displaced in some degree (for a general account of this problem, see 13)). For example, it was noticed[14)-17)] that the ground state of *heptalene* does not show an energy minimum for the nuclear configuration with D_{2h} symmetry, suggested by the superposition of the two Kekulé-type resonance structures, but has a lower energy if it adopts a less symmetrical nuclear configuration that corresponds to either of the resonance structures and exhibits a strong bond-length alternation in its periphery. The resonance between the Kekulé-type structures in this molecule would substantially be hindered.

The available experimental facts[6)] agree with this in indicating that the π-electrons in this molecule should be localized largely in "double" bonds, rather than uniformly delocalized over the entire molecule.

Another example is provided by [30] *annulene*. Longuet-Higgins and Salem[18] have shown that the observed visible and UV absorption spectrum and, in particular, the NMR proton chemical shifts of this molecule are very difficult to reconcile with the symmetrical nuclear configuration (D_{6h}) suggested by the superposition of the Kekulé-type resonance structures. The hypothesis of a bond-length alternation of D_{3h} symmetry removes this difficulty. This indicates that the resonance between Kekulé-type structures should be very much impeded also in this molecule.

A theoretical explanation for such an anomalous phenomenon in certain nonalternant hydrocarbons has first been attempted, in case of pentalene, by Boer-Veenendaal and Boer[19], followed by Boer-Veenendaal et al.[14], Snyder[15], and Nakajima and Katagiri[17] for other related nonalternant hydrocarbons. By making allowance for the effects of σ-bond compression, these authors have shown that a distorted structure resembling either of the two Kekulé-type structures is actually energetically favored as compared with the apparently-full symmetrical one.

A theoretical justification for the existence of a bond-length alternation accompanied by a molecular-symmetry reduction in higher members of the [4n + 2] annulenes came from a somewhat different source. It is Kuhn[20] who first showed that in contradiction to the earlier predictions by Lennard-Jones[21] and Coulson[22], a certain degree of bond alternation should be postulated even in an infinitely-long linear polyene, the assumption providing a possible theoretical interpretation for the experimental fact that in the electronic spectra of chain polyenes, the frequency of the longest wave-length absorption band converges to a finite limit, as the chain length tends to infinity, rather than to zero, as is expected on the basis of the conventional molecular-orbital theory. Since an infinite chain polyene is not to be distinguished from an infinite cyclic polyene, this result by Kuhn implies at once that some degree of bond alternation is also present in a sufficiently large cyclic polyene, even if it satisfies the Hückel 4n + 2 rule. Such a view has been followed up by succeeding theoretical works by Dewar[23], Simpson[24], Platt[25], Huzinaga and Hashino[26], Labhart[27], Ooshika[28], Longuet-Higgins and Salem[29], Coulson and Dixon[30], and Förstering et al.[31]. Simpson[24] has tried to rationalize the experimental slow convergency of the absorption frequency in chain polyenes by treating the "double" bonds as independent ethylenic systems and even by neglecting the exchange interaction between adjacent systems. It was emphasized by Labhart[27] and Kuhn[20] that the stable nuclear configuration of a conjugated system cannot be predicted on the basis of π-electron calculations alone, but allowance for the effects of σ-bond compression has to be made. It is Platt[25] who first suggested that the stabilization of the distorted structure in a long chain polyene is due to the *vibronic interaction* of the ground state with low-lying excited states of proper symmetry—that is, the pseudo Jahn-Teller effect. As for the critical value of n for which bond alternation sets in in $[4n+2]$ annulenes, a variety of values ranging from 2 to 8 have been proposed: $n=2$ by Coulson and Dixon[30] and Nakajima[32], $n=5$ by Dewar and Gleicher[33], Binsch and Heilbronner[34], and Binsch et al.[35], $n>5$ by Förstering et al.[31], and $n=8$ by Longuet-Higgings and Salem[29], who proposed alternatively a value between 4 and 7[18].

A basic assumption common to all these theoretical treatments is that a bond alternation corresponding to one of the Kekulé-type structures is the energetically most favorable bond distortion in a conjugated molecule. Even if this be so with the ground states of simple conjugated molecules (e. g., pentalene), obviously, such a presumption cannot be extended to large polycyclic conjugated molecules (e. g., Hafner's hydrocarbons) in which possible Kekulé-type resonance structures are not always equivalent. For the same reason it cannot be applied to the charged conjugated systems or to the electronically excited states. Thus, a reexamination of the theory of double-bond fixation is highly desirable, since, in an advanced theory, not only bond alternation but all the possible types of bond distortion should be examined.

Recently, Binsch et al.[34]–[38] and Nakajima et al.[39]–[42] have, along this line, developed a *general theory of double-bond fixation in conjugated molecules*. The scheme of Binsch is based on the *second-order perturbation theory* and allows a sharp distinction to be made between the first-order bond fixation, which does not affect the molecular symmetry, and the second-order bond fixation which may result in a molecular-symmetry reduction. Information about the second-order bond fixation is obtained by examining the eigenvalues and eigenvectors of the bond-bond polarizability matrix. If the largest eigenvalue of the bond-bond polarizability matrix for a molecule is larger than a certain critical value, the second-order effects in the π-electron energy overcome the σ-bond compression energy, and the molecule will, in general, distort into a less symmetrical nuclear configuration.

The type of the most favorable second-order bond distortion is given by the eigenvector corresponding to the largest eigenvalue.

On the other hand, Nakajima et al.[41), 42)] have applied the *symmetry rule* recently developed by Pearson[44)] to the estimation of the second-order bond distortion. This rule provides a more intelligible way of predicting the molecular-symmetry reduction occurring in certain conjugated molecules and of understanding its origin.

Since they are based on the perturbation theory, both these theories—so to speak, the static theories—only give the type of the most favorable bond fixation and do not provide the actual magnitude of bond distortion or equilibrium bond distances at which the nuclei of the real molecule will settle. A general theory for predicting the *energetically most favorable geometrical structure* with respect to bond distance of a conjugated molecule—a dynamic theory—has been developed by Nakajima and Toyota[39), 40)]. The method used is the semi-empirical SCF MO method in conjunction with the variable bond-length technique. By adopting all the possible distorted structures as the starting geometries for iterative calculation, we can obtain automatically the energetically most favorable nuclear arrangement using this method.

In this contribution, we are concerned with the static and dynamic theories—in a sense mentioned above—of bond distortion in conjugated hydrocarbons. The geometrical structures of the ground states of non-alternant hydrocarbons, some of their charged radicals and dianions, and $[4n+2]$ annulenes, together with those of electronically excited states of selected molecules will be treated. Further, the effects of bond distortion on the magnetic susceptibilities and electronic excitation energies, the physical quantities which depend sensitively upon geometrical structure, in nonalternant hydrocarbons will be discussed.

II. The Static Theory of Bond Distortion

A. The Second-Order Perturbation Theory

We start by assuming for a conjugated molecule a fully-symmetrical arrangement of carbon nuclei as an unperturbed system. Electronic wavefunctions $\psi_0, \psi_1, \ldots, \psi_n, \ldots$ and the corresponding energies $E_0, E_1, \ldots, E_n, \ldots$ of the unperturbed system are assumed to be known. We then distort the nuclei from the symmetrical arrangement by means of the ith normal coordinate of nuclear motion Q_i. So long as the distortion is not too drastic, we may describe the ground-state energy after the deformation using the second-order perturbation theory:

$$E(Q_i) = E_0 + \left\langle \psi_0 \left| \frac{\partial H}{\partial Q_i} \right| \psi_0 \right\rangle Q_i$$

$$+ \frac{1}{2} \left\{ \left\langle \psi_0 \left| \frac{\partial^2 H}{\partial Q_i^2} \right| \psi_0 \right\rangle - 2 \sum_{n \neq 0} \frac{\left| \left\langle \psi_n \left| \frac{\partial H}{\partial Q_i} \right| \psi_0 \right\rangle \right|^2}{E_n - E_0} \right\} Q_i^2 . \tag{1}$$

Now we assume the complete $\sigma - \pi$ separation for a conjugated molecule, which states that the total Hamiltonian can be written as the sum of the Hamiltonian for the σ-core and that for the π-electron system:

$$H = H_\sigma + H_\pi . \tag{2}$$

Eq. (2) means further that the total energy can correspondingly be written as the sum of the two terms:

$$E = E_\sigma + E_\pi . \tag{3}$$

The σ-core energy can be regarded as the sum of the individual contributions of the C—C σ-bonds, each of which may be approximated by a quadratic function of the bond-distance variation:

$$E_\sigma = \sum_{\mu < \nu} \frac{k}{2} (r_{\mu\nu} - r_0)^2 \tag{4}$$

where k is the force constant appropriate for an sp^2 hybridized carbon σ-bond.

If the initial ground-state wavefunction ψ_0 is nondegenerate, the first-order term (i. e., the second term) in Eq. (1) is nonzero only for the totally-symmetrical nuclear displacements (note that Q_i and $(\partial H / \partial Q_i)$ have the same symmetry). Information about the equilibrium nuclear configuration after the symmetrical first-order deformation will be given by equating the first-order term to zero,

$$\left\langle \psi_0 \left| \frac{\partial H}{\partial Q_i} \right| \psi_0 \right\rangle = \frac{\partial E_0}{\partial Q_i} = 0 \tag{5}$$

where for the first equality we have used the Hellman-Feynmann theorem[43]. Further, Eq. (5) can be rewritten as

$$\frac{\partial E_0}{\partial Q_i} = \sum_{\mu < \nu} \left(\frac{\partial E_0}{\partial r_{\mu\nu}} \right) \left(\frac{\partial r_{\mu\nu}}{\partial Q_i} \right) = 0 \tag{6}$$

T. Nakajima

which means that all the partial derivatives of the total energy with respect to $r_{\mu\nu}$ should vanish individually. Writing the π-electron energy in the one-electron approximation and using Eqs. (3) and (4), we have for the total energy

$$E_0 = \sum_{\mu<\nu} \left\{ \frac{k}{2}(r_{\mu\nu}-r_0)^2 + 2p_{\mu\nu}\beta_{\mu\nu} \right\} + N\alpha \tag{7}$$

where $p_{\mu\nu}$ and $\beta_{\mu\nu}$ are respectively the bond order and the resonance integral for real bond $\mu-\nu$, and α is the Coulomb integral for the carbon atom. Differentiation of Eq. (7) with respect to $r_{\mu\nu}$ yields the following relations for all bonds

$$r_{\mu\nu}^{(1)} = r_0 - \frac{2\beta'}{k}p_{\mu\nu}. \tag{8}$$

Eq. (8) is nothing but the usual bond-order—bond-length relationship for conjugated hydrocarbons. It follows that in these molecules all the symmetrical bond distortions will occur until the first-order energy equilibrium is attained. In other words, initially assumed bond lengths will change to the first-order equilibrium values through the bond-order—bond-length relationship, but still keep the initial molecular-symmetry.

Assuming that the first-order changes have taken place and using Eqs. (2), (3) and (4), we find

$$E(Q_i) = E_0^{(1)} + \frac{1}{2}\left\{ k + \left\langle \psi_0 \left| \frac{\partial^2 H_\pi}{\partial Q_i^2} \right| \psi_0 \right\rangle - 2 \sum_{n\neq 0} \frac{\left| \left\langle \psi_n \left| \frac{\partial H_\pi}{\partial Q_i} \right| \psi_0 \right\rangle \right|^2}{E_n - E_0} \right\} Q_i^2. \tag{9}$$

In the framework of the one-electron approximation, the second term in braces in Eq. (9) can be written as

$$\left\langle \psi_0 \left| \frac{\partial^2 H_\pi}{\partial Q_i^2} \right| \psi_0 \right\rangle = 2 \sum_{\mu<\nu} p_{\mu\nu}\beta_{\mu\nu}'' \left(\frac{\partial r_{\mu\nu}}{\partial Q_i} \right)^2 \tag{10}$$

where β'' is the second derivative of $\beta(r)$ with respect to r. We may safely neglect this term since for the reduced bond-distance interval in question the curvature of the $\beta(r)$ curve would be very small. We have

8

finally for the energy of the ground state after the second-order deformation

$$E(Q_i) = E_0^{(1)} + \frac{1}{2}\left\{k - 2\sum_{n \neq 0} \frac{\left|\left\langle \psi_n \left| \frac{\partial H_\pi}{\partial Q_i} \right| \psi_0 \right\rangle\right|^2}{E_n - E_0}\right\} Q_i^2.$$ (11)

According to Eq. (11), the force constant for the normal vibration Q_i can be identified with the term in braces and can be *negative* if the second term, which is positive, exceeds the first term. If the force constant is negative, the energy should be lowered by the nuclear deformation Q_i, and the second-order distortion from the symmetrical nuclear arrangement would occur spontaneously.

Two different methods for estimating the probable value of force constant have so far been proposed; in estimating the second term in braces in Eq. (11), Binsch et al.[34)–38)] have used the bond-bond polarizabilities, while Nakajima et al.[41), 42)] have made an extensive use of the symmetry rule, the usefulness of which in predicting stable molecular shapes has recently been appreciated by Pearson[44), 45)].

B. Use of Bond-Bond Polarizabilities

In the framework of the one-electron approximation, the second term in braces of Eq. (11) is given by

$$2\sum_{n \neq 0} \frac{\left|\left\langle \psi_n \left| \frac{\partial H_\pi}{\partial Q_i} \right| \psi_0 \right\rangle\right|^2}{E_n - E_0} = -\sum_{\mu < \nu}\sum_{\kappa < \lambda} 2\pi\,_{\mu\nu,\kappa\lambda}^{(1)}\,\beta'_{\mu\nu}\beta'_{\kappa\lambda}\left(\frac{\partial r_{\mu\nu}}{\partial Q_i}\right)\left(\frac{\partial r_{\kappa\lambda}}{\partial Q_i}\right)$$ (12)

where $\pi_{\mu\nu,\kappa\lambda}$ is the bond-bond polarizability, and we have neglected the curvature of the $\beta(r)$ curve (i. e., we have assumed $\beta''=0$). Since Eq. (12) contains cross-terms, it is convenient to apply a normal-coordinate analysis. Diagonalization of the bond-bond polarizability matrix will yield a set of eigenvalues λ_i and the corresponding set of eigenvectors D_i. We can then rewrite Eq. (11) as

$$E(Q_i) = E_0^{(1)} + \tfrac{1}{2}\{k + 2\lambda_i(\beta')^2\}Q_i^2.$$ (13)

Remembering that the polarizabilities are expressed in units of β_0^{-1}, it follows that the largest positive eigenvalue λ_{max} corresponds to the energetically most favorable second-order bond distortion, and the type

9

of this distortion is given by the eigenvector D_{max} belonging to λ_{max}. If the largest eigenvalue λ_{max} exceeds a certain critical value λ_{crit}, which will be given by

$$\lambda_{crit} = -\frac{k\beta_0}{2(\beta')^2} \qquad (14)$$

then the force constant for the corresponding normal vibration will be negative, and the molecule is predicted to suffer the second-order bond distortion which is, in general, accompanied with a molecular-symmetry reduction. The type of symmetry reduction is given by the irreducible representation to which D_{max} and λ_{max} belong, relative to the point group of the initial, fully symmetrical nuclear configuration. Using the reasonable parameter values for β_0, k and $(2\beta'/k)$ which appears in Eq. (8), Binsch estimated $\lambda_{crit} = 1.81\,\beta_0^{-1}$ for use in the Hückel model[34] and $\lambda_{crit} = 1.22\,\beta_0^{-1}$ for use in the semiempirical SCF MO formalism[35, 38].

Using the above procedure and the λ_{crit} values, Binsch has examined the second-order bond fixations in the ground states of linear, cyclic, and benzenoid hydrocarbons[34, 35, 36], and nonalternant hydrocarbons and nitrogen heterocycles[37, 38], together with those in the lowest excited states of nonalternant hydrocarbons[37].

C. Application of the Symmetry Rule

Recently, a symmetry rule for predicting stable molecular shapes has been developed by Pearson[44, 45], Salem[46], and Bartell[47]. This rule is based on the second-order, or pseudo, Jahn-Teller effect and follows from the earlier work by Bader[48, 49]. According to the symmetry rule, the symmetries of the ground state and the lowest excited state determine which kind of nuclear motion occurs most easily in the ground state of a molecule. Pearson has shown that this approximation is justified in a large variety of inorganic and small organic molecules.

In estimating the probable value of the force constant, we now make use of this approximation and assume that the infinite sum over the excited states in Eq. (11) is replaced by one or two dominant terms corresponding to the lowest one or two excited states. Our approach is then simply to examine whether a given molecule in a symmetrical nuclear configuration has reasonably low first excited state(s) and, if so, whether any of the matrix elements $\langle \psi_n | \partial H_n / \partial Q_i | \psi_0 \rangle$ are nonvanishing. Since Q_i and $(\partial H_n / \partial Q_i)$ have the same symmetry and since the ground state is, in general, totally symmetric, the integral is nonzero only when ψ_n and Q_i have the same symmetry. The symmetry of the lowest excited state(s) now determines which kind of nuclear displacement occurs energetically most easily.

Although the symmetry of the most soft nuclear distortion can thus be determined, its actual type cannot always be uniquely determined, because there are, in general, several types of bond distortion belonging to the same symmetry. In order to predict which type of distortion is actually energetically most favorable, it is useful to interpret the second term in braces of Eq. (11) as the "relaxability" of the molecule along the coordinate Q_i, and to express the matrix element $\langle \psi_n | \partial H_\pi / \partial Q_i | \psi_0 \rangle$ in terms of the transition density ρ_{0n} between the ground and excited states[46]:

$$\left\langle \psi_n \left| \frac{\partial H_\pi}{\partial Q_i} \right| \psi_0 \right\rangle = \int \rho_{0n} \frac{\partial v}{\partial Q_i} \, d\tau \qquad (15)$$

where v is the one-electron nuclear-electron potential operator. For a one-electron transition between molecular orbitals φ_i and φ_j, ρ_{0n} is given by $\sqrt{2} \, \varphi_i \varphi_j$[13]. A given excited state may contribute much to the molecular relaxability towards mode Q_i, if ρ_{0n} is such that the distribution of its one-center and two-center components matches with that of the components of displacement Q_i, but little to the molecular relaxability, if the distribution of components of ρ_{0n} does not correspond to that of Q_i. We can thus predict the actual type of the most soft bond distortion by examining the distribution of the components of transition density over the entire molecular skeleton.

D. Results and Discussion

The λ_{max} values and the symmetry types of the corresponding bond distortions for the ground states of nonalternant hydrocarbons (Fig. 5) are listed in Table 1. Two sets of λ_{max} values are presented: one set of values obtained from the bond-bond polarizabilities calculated using the Hückel MO method[37] and the other set obtained from those calculated using the semiempirical SCF MO method[38]. Also shown in Table 1 are the energies and symmetries of the lowest excited states obtained using the semiempirical SCF CI MO method in conjunction with the variable bond-length technique (vide infra) assuming the full molecular symmetries for the ground states. Further, in the last column of Table 1 are presented the types of molecular-symmetry reduction predicted using the dynamic theory (vide infra) for molecules in which it occurs.

The second-order bond distortions predicted on the basis of the Hückel λ_{max} values are not in good agreement with those predicted using the dynamic theory: according to results based on the dynamic theory, of the five molecules, I, IV, V, VII and X, λ_{max} values of which are larger than the critical value $(1.8\,\beta_0^{-1})$, molecules IV and X do not suffer a second-order bond distortion, and three molecules, III, IX and XIII, λ_{max} values of which are all smaller than the critical value do suffer a symmetry reduction. The second-order bond distortions predicted from the SCF λ_{max} values are in much better agreement with those predicted using the dynamic theory: only molecule V which is predicted to have a λ_{max} value smaller than the critical value $(1.22\,\beta_0^{-1})$, but to suffer a symmetry reduction according to the dynamic theory, is an exceptional case.

We now move to the predictions based on the symmetry rule. Comparing the lowest excitation energies (the sixth column) with the results concerning the symmetry reduction obtained on the basis of the dynamic theory (the last column), we can draw a very clear-cut criterion for the symmetry reduction:

> if the lowest excitation energy of a given molecule, calculated assuming the full molecular symmetry, is *smaller* than ca. 1.2 eV, the force constant for certain antisymmetrical in-plane nuclear vibration should be *negative*, and the molecule would be distorted into a less symmetrical nuclear arrangement.

The proposed criterion holds for all the molecules without exceptions. It is interesting to note that in estimating the probable value of force constant the replacement of the sum over excited states (the second term in braces of Eq. (11)) more simply by the term corresponding to the lowest excited state brings about the results which agree quite well with those obtained using a far more sophisticated method. As will be shown below, for molecules which suffer a molecular-symmetry reduction, there is indeed a good correlation between the lowest excitation energies and the relative stabilities of the distorted structures, i. e., the stabilization energies defined as the difference in total energy between the hypothetical fully-symmetrical structure and the actual distorted one *(vide infra)*. A plot of the SCF λ_{max} values against stabilization energies shows a far less successful correlation.

The symmetries of the lowest excited states listed in Table 1 are nothing but the symmetries to which the most soft second-order bond distortions belong. It is seen that the types of symmetry reduction predicted using the symmetry rule are in complete agreement with those obtained on the basis of the dynamic theory.

In Fig. 1 are shown the one-center and two-center components of transition densities ρ_{01} for some selected molecules. From these data the actual types of the most soft nuclear displacements will be deter-

Table 1. The second-order bond distortion in the ground states of nonalternant hydrocarbons

Molecule[a] (symmetry)[b]	Hückel[c]		SCF[d]		Lowest excited state		Symmetry reduction[f]
	λ_{max} (β_0^{-1})	Type of distortion	λ_{max} (β_0^{-1})	Type of distortion	$E_1 - E_0$ (eV)[e]	Symmetry	
I (D_{2h})	2.36	B_{3g}	3.15	B_{3g}	0.35	B_{3g}	$D_{2h} \rightarrow C_{2h}$
II (D_{2h})			1.83	B_{1u}	1.22	B_{1u}	$D_{2h} \rightarrow C_{2v}$
III (D_{2h})	1.57	B_{3g}	1.43	B_{3g}	1.00	B_{3g}	$D_{2h} \rightarrow C_{2h}$
IV (C_{2v})	2.06	B_2	0.65	A_1	1.47	B_2	
V (C_{2v})	1.83	B_2	1.13	B_2	0.41	B_2	$C_{2v} \rightarrow C_s$
VI (C_{2h})	1.01	A_g	0.66	A_g	2.54	A_g	
VII (D_{2h})	2.60	B_{3g}	2.59	B_{3g}	0.26	B_{3g}	$D_{2h} \rightarrow C_{2h}$
VIII (D_{2h})					0.81	B_{1u}	$D_{2h} \rightarrow C_{2v}$
IX (D_{2h})	1.75	B_{3g}	1.38	B_{3g}	0.83	B_{3g}	$D_{2h} \rightarrow C_{2h}$
X (C_{2v})	2.82	B_2			1.46	B_2	
XI (C_{2v})	1.26	B_2	1.11	B_2	2.05	B_2	
XII (C_{2v})					1.07	B_2	$C_{2v} \rightarrow C_s$
XIII (C_{2v})	1.34	B_2	1.32	B_2	0.81	B_2	$C_{2v} \rightarrow C_s$
XIV (C_{2v})	1.14	B_2	1.11	B_2	1.73	B_2	
XV (C_{2v})	1.39	B_2	1.11	B_2	1.52	B_2	
XVI (D_{2h})			0.69		2.24	B_{3g}	
XVII (C_{2v})	1.17	A_1	0.68		2.98	A_1	
XVIII (D_{2h})					2.15	B_{3g}	
XIX (C_{2v})					3.32	B_2	
XX (C_{2v})					2.97	B_2	
XXI (D_{2h})					2.39	B_{3g}	
XXII (C_{2v})					2.72	B_2	
XXIII (D_{2h})					2.14	B_{3g}	
I^{-2} (D_{2h})			0.47		4.62	B_{2u}	
III^{-2} (D_{2h})			0.62		3.05	B_{2u}	
IV^{-2} (C_{2v})			0.52		3.93	B_2	
VII^{-2} (D_{2h})					1.43	B_{2u}	

[a] See Fig. 5.
[b] The apparently-full molecular symmetry.
[c] In this approximation $\lambda_{crit} = 1.8 \beta_0^{-1}$.
[d] In this approximation $\lambda_{crit} = 1.22 \beta_0^{-1}$.
[e] $(E_1 - E_0)_{crit} \simeq 1.2$ eV.
[f] Predicted using the dynamic theory.

mined. For example, the symmetry of the most soft nuclear displacement for pentalene (I) is predicted to be b_{3g}. However, in this molecule two distinct types of bond distortion are possible. The components of the transition densities for pentalene shown in Fig. 1 indicate that the lowest excited state contributes much to the molecular relaxability towards

13

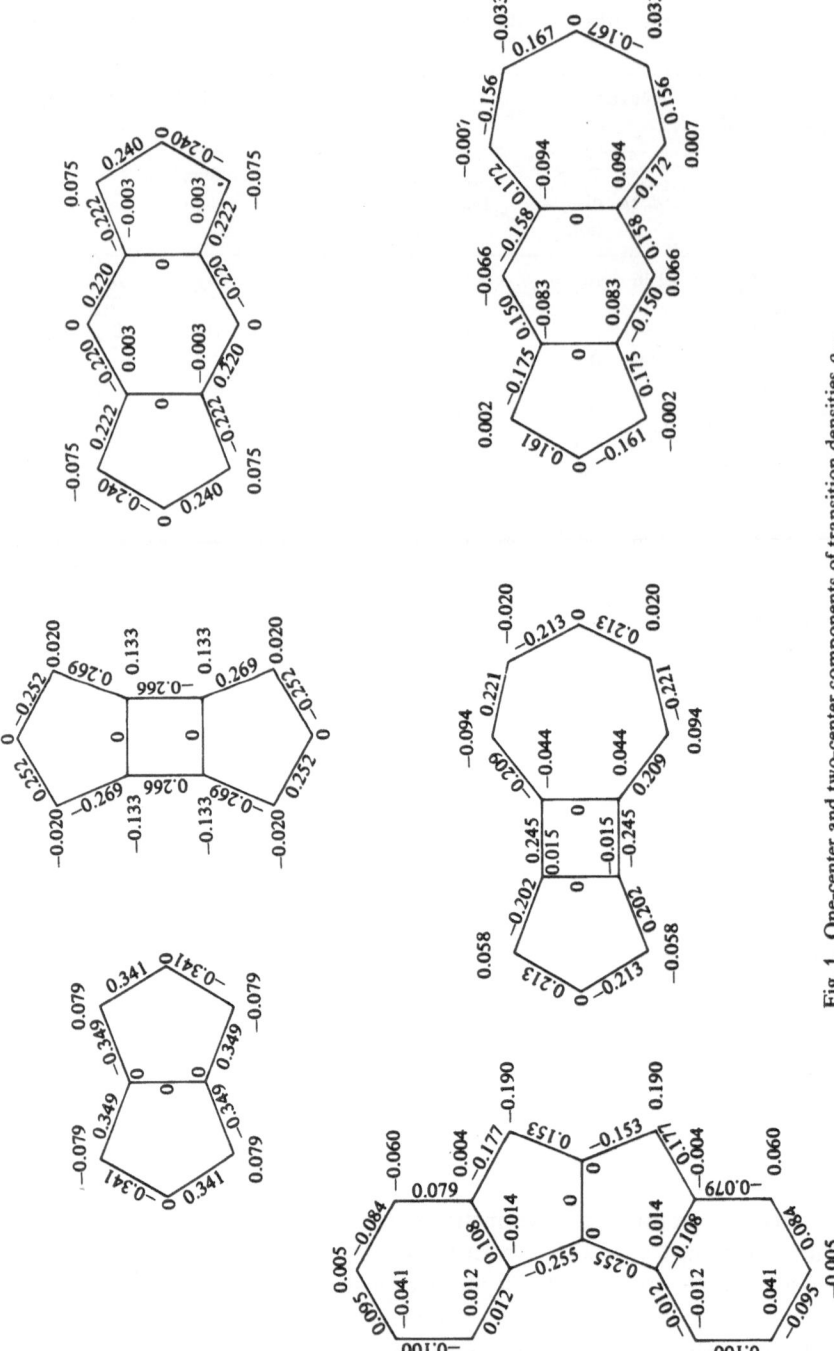

Fig. 1. One-center and two-center components of transition densities ρ_{10}

the bond-alternation type of distortion. Other data in Fig. 1 show that in most of the molecules examined, with the exception of molecule V, the bond alternation in the molecular periphery is energetically most favored. Such will be the case with the 7-membered analogues of molecules I, II and III. As will be seen later, the bond-length distributions calculated using the dynamic theory support these predictions.

The molecular-symmetry reductions in molecules I, III, V, VII, IX and XII are closely related to the fact that their peripheral skeletons correspond to $4n$ cyclic polyenes. In the Hückel picture the top half filled molecular orbitals of $4n$ cyclic polyenes are doubly degenerate, and the introduction of a transannular bond to form a nonalternant hydrocarbon lifts the degeneracy. Orbital splittings are in general rather small, and the resultant nonalternant hydrocarbons have still a very low-lying excited state. Fig. 2 shows how the degenerate nonbonding orbitals of cyclooctatetraene split by the introduction of a transannular bond to form pentalene. The changes in orbital energy have been evaluated using the first-order perturbation theory:

$$\delta\varepsilon_i = 2C_{\mu i}C_{\nu i}\beta \qquad (16)$$

where C_μ and C_ν are the atomic-orbital coefficients for atoms μ and ν between which a transannular bond is introduced. One of the nonbonding orbitals has its energy lowered after the perturbation, while the other has its energy unchanged. The resultant pentalene possesses still a nonbonding orbital which now is empty and, consequently, has a fairly low-lying excited state which can interact effectively with the ground state if a relevant nuclear deformation is applied. The type of the bond distortion which makes this interaction possible is given by examining the transition density ρ_{01}. Since the perturbation due to the introduction of a transannular bond is not so strong, the transition density, $\sqrt{2}\varphi$ (top filled) $\times \varphi$ (bottom empty), for pentalene may be approximated by $\sqrt{2}\varphi$ (one nonbonding orbital) $\times \varphi$ (the other nonbonding orbital) for cyclooctatetraene. Using the atomic-orbital coefficients for nonbonding orbitals shown in Fig. 2, we can easily show that the distribution of one-center and two-center components of the transition density between nonbonding orbitals is such that it favors the bond alternation.

It should be noted, on the other hand, that a symmetry reduction is predicted even in molecules II, VIII and XIII whose peripheral skeletons correspond to $4n+2$ cyclic polyenes. The transannular bonds in these molecules are different in nature from those mentioned above. For example, the introduction of the transannular bonds between atoms 2 and 8 and between 3 and 7 of cyclododecapentaene to form bowtiene (II) (Fig. 3) brings about the splitting of the top filled degenerate orbitals of the unperturbed system into two levels, one with its energy raised and

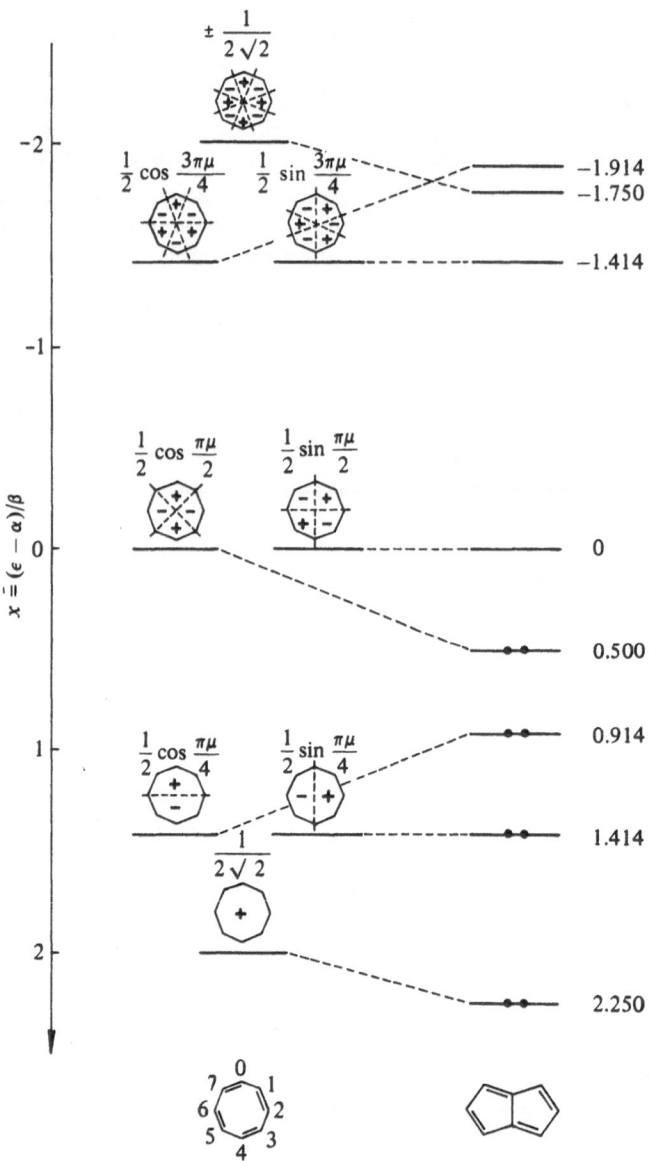

Fig. 2. Correlation of the molecular orbitals of cyclooctatetraene with those of pentalene

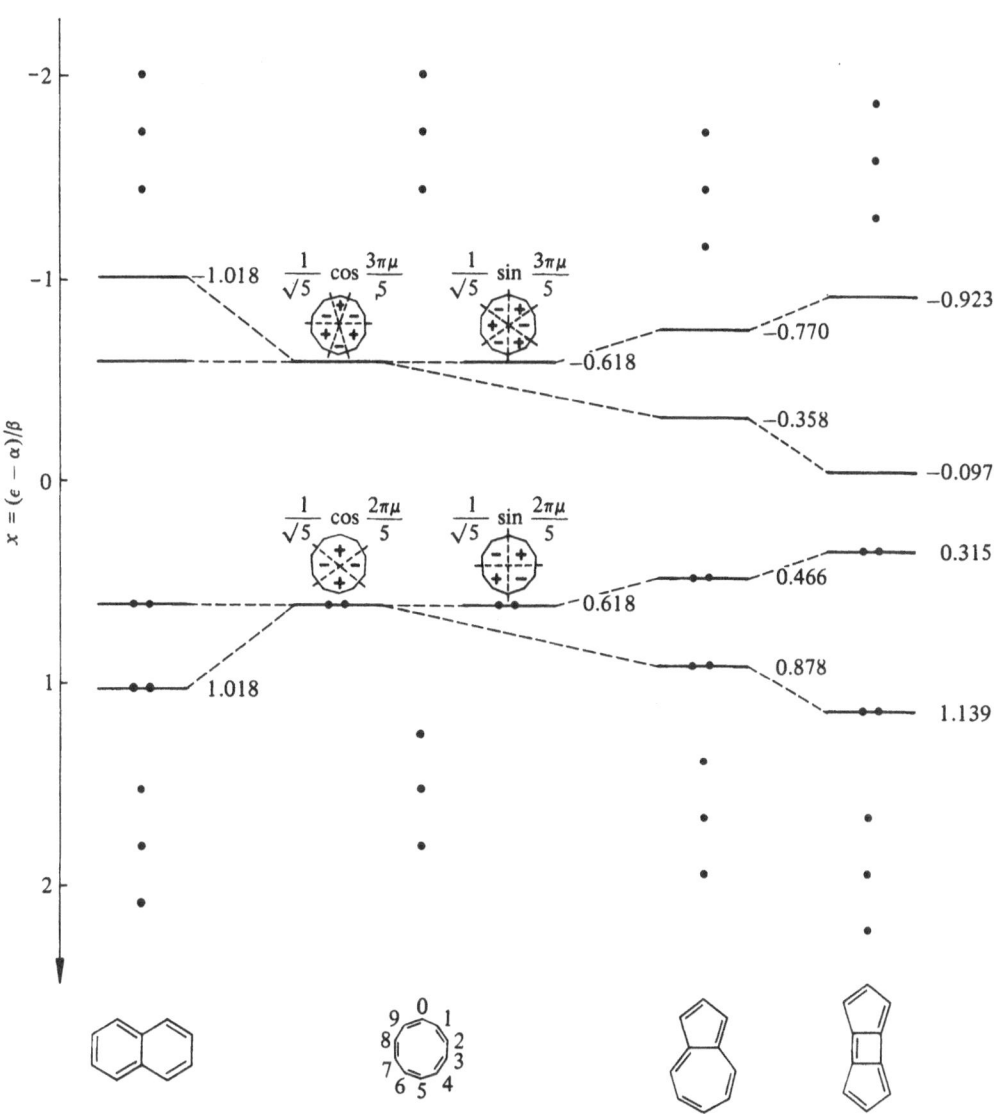

Fig. 3. Correlation of the orbitals of cyclododecapentaene with those of azulene, bowtiene and naphthalene

the other with its energy lowered and, at the same time, a similar splitting of the bottom empty degenerate orbitals. As a result, in the resultant bowtiene a very low-lying excited state appears which can interact effectively with the ground state, if the molecule suffers a relevant nuclear deformation. The type of bond distortion which makes such an inter-action possible may be conjectured from the product of molecular or-bitals of the unperturbed system with which the top filled and bottom empty orbitals of bowtiene are correlated. Such orbitals of cyclododecapentaene are the highest filled orbital with atomic-orbital coefficients $(1/\sqrt{5})\sin(2\pi\mu/5)$ $(\mu=0, 1, \ldots 9)$ and the lowest empty orbital with atomic-orbital coefficients $(1/\sqrt{5})\cos(3\pi\mu/5)$ $(\mu=0,1,\ldots 9)$, and the product of these orbitals gives the transition density which indicates that the lowest excited state of bowtiene contributes much to the molecular relaxability towards the bond-alternation type of bond distortion.

Azulene (XI) possesses a transannular bond which has the same effect as those of bowtiene (Fig. 3). The splittings of the top filled and bottom empty degenerate orbitals of cyclododecapentaene in this case are half the corresponding splitting in the case of bowtiene, and are not large enough to produce an effective vibronic interaction between the ground and lowest excited states of the resultant azulene molecule.

Fig. 3 also includes the splittings of the top filled and bottom empty degenerate orbitals of cyclododecapentaene by the introduction of a transannular bond between atoms 0 and 5 to produce naphthalene. The situation is essentially different from that observed in the case of bowtiene or azulene: one of the top filled degenerate orbitals remains unchanged in energy by the perturbation and, likewise, one of the bottom empty degenerated orbitals undergoes no energy change, the lowest electronic transition in naphthalene occurring between these orbitals. The lowest excited state of naphthalene is too high to interact effectively with the ground state, and no symmetry reduction from the fully-symmetrical nuclear configuration (D_{2h}) is expected.

In Table 1 the dianions of some selected nonalternant hydrocarbons are included. All the dianions examined have the lowest excitation energies larger than the critical value (ca. 1.2 eV), and are predicted to undergo no symmetry reduction in agreement with the results obtained using the dynamic theory. The orbital arrangement for pentalene shown in Fig. 2 serves for showing how high the lowest excited state is when two extra electrons are placed in the nonbonding orbital to form the dianion.

Next, we consider the application of the symmetry rule to the pre-diction of the geometrical structures of some other related systems.

Recently the anion and cation radicals of heptafulvalene (XXIII) have been prepared and their ESR spectra have been investigated by

Sevilla *et al.*[50]. The analysis of the hyperfine spectra of these radicals has revealed that the cation radical should have D_{2h} symmetry, whereas the molecular symmetry of the anion radical should be lower than the apparently-full molecular symmetry, i. e., D_{2h}. We now examine the molecular symmetries of the anion and cation radicals of heptafulvalene, together with those of its 5-membered analogue, fulvalene (XXI) using the symmetry rule.

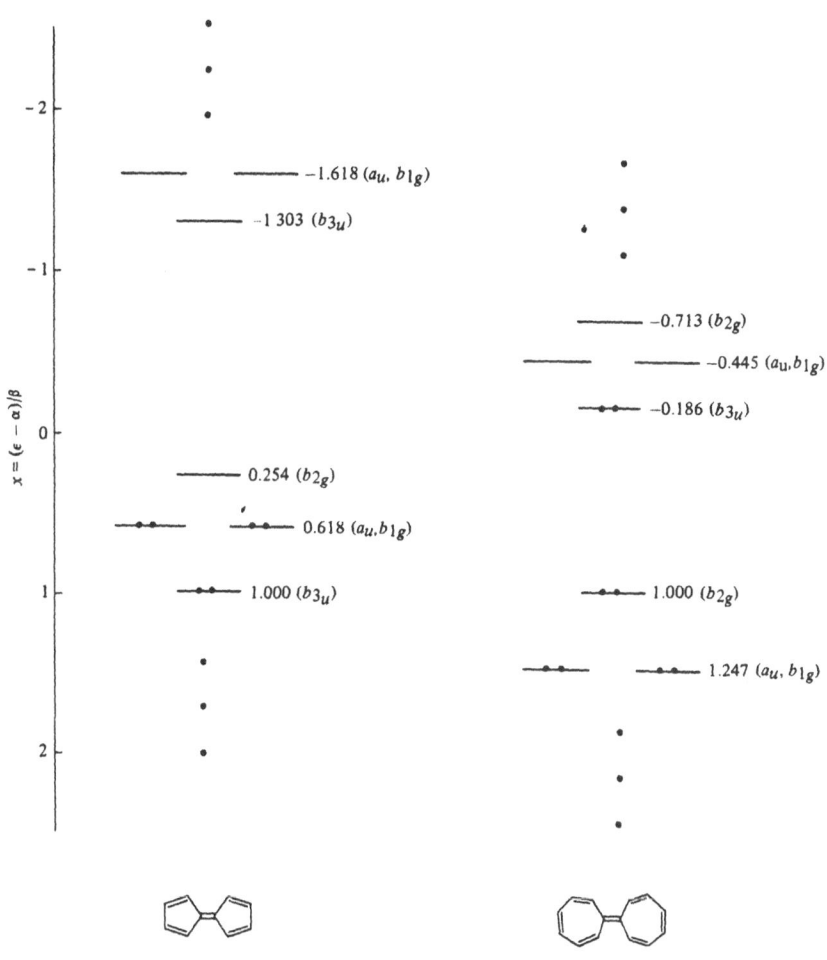

Fig. 4. Hückel molecular orbitals of fulvalene and heptafulvalene

If the full molecular symmetry (D_{2h}) is assumed, the ground states of the cation radical of fulvalene and the anion radical of heptafulvalene are both predicted to be of B_{1g} symmetry[52] by using the semiempirical open-shell SCF MO method[51]. The lowest excited states of both radicals are of A_u symmetry and are predicted to be very close to the ground states: in the framework of the Hückel approximation these states are degenerate in both cases (Fig. 4). Therefore, it is expected that in both these radicals the ground state interacts strongly with the lowest excited state through the nuclear deformation of symmetry $b_{1u}(z)$ $(= a_u \times b_{1g})$, with the result that the initially assumed molecular symmetry (D_{2h}) should be reduced to C_{2v}.

On the other hand, in the anion radical of fulvalene and the cation radical of heptafulvalene, the energy gaps between the ground and lowest excited state (which is in both cases doubly degenerate in the Hückel approximation (Fig. 4)) are predicted to be reasonably large (1.4 and 1.7 eV, respectively), so that these radicals would not suffer a symmetry reduction.

It will be seen below that the actual geometrical structures of these radicals calculated using the dynamic theory support these conclusions.

Next, we discuss the *possibility of molecular-symmetry reduction in annulenes*. It has been noticed by several authors[13), 15), 53)−56)] that cyclobutadiene, the simplest $4n$ annulene, should take a rectangular form rather than the fully-symmetrical, square form in the ground state. In the Hückel picture it possesses a pair of degenerate nonbonding orbitals in which the top two electrons are accomodated. There are four ways of assigning two electrons to these orbitals, and in the simplest one-electron model the three singlet states ($^1A_{1g}$, $^1B_{1g}$ and $^1B_{2g}$, if the z axis is chosen perpendicular to the molecular plane) and the triplet state (^3A_g) are all degenerate. Inclusion of electron interaction lifts the degeneracy and gives rise to four distinct states separated by small energy differences: the triplet state and the $^1A_{1g}$, $^1B_{1g}$ and $^1B_{2g}$ singlet states in order of increasing energy[13), 15)]. The lowest singlet can, in principle, interact vibronically with the $^1B_{1g}$ and $^1B_{2g}$ singlets through the b_{1g} and b_{2g} deformations of the nuclei, respectively. Of these two possible types of nuclear deformation, the b_{2g} distortion, which corresponds to an antisymmetric C—C stretching mode (a bond alternation), would probably be soft as compared with the b_{1g} distortion, which includes the bond-angle deformation. The energy of the $^1A_{1g}$ state is thus lowered by the vibronic interaction with the $^1B_{2g}$ state to such an extent that it finally lies below the triplet state. The ground state of cyclobutadiene should, therefore, be a rectangular singlet rather than a square triplet. A quite similar phenomenon is expected to occur in all the higher homologues of $4n$ annulenes.

On the other hand, the situation is rather different in $[4n+2]$ annulenes. *Benzene* in its fully-symmetrical nuclear arrangement (D_{6h}) has the totally symmetric ground state $({}^1A_{1g})$ and the lower excited singlet states of ${}^1B_{2u}$, ${}^1B_{1u}$ and ${}^1E_{1u}$ symmetries (in order of increasing energy). In principle, the vibronic interaction between the ground state and the lowest-lying excited singlet $({}^1B_{2u})$ is possible through the antisymmetric nuclear deformation of b_{2u} symmetry, which corresponds to the bond alternation. However, since the energy gap between the two states is large (ca. 5 eV), the vibronic interaction is not large enough to make the corresponding force constant negative and to distort the ground state into a skewed structure of D_{3h} symmetry. This argument provides a very comprehensive way of understanding the reason why benzene adopts a regular-hexagonal nuclear arrangement in its ground state, and, at the same time, suggests that such a vibronic coupling will be large enough to distort the ground state into a less symmetrical nuclear configuration, as soon as the lowest excitation energy becomes small enough in the large $[4n+2]$ annulenes.

For $C_{18}H_{18}$ and $C_{30}H_{30}$, both known to date[11), 12)], the excitation energies for the lowest ${}^1A_{1g} \rightarrow {}^1B_{2u}$ transition, calculated assuming molecular symmetry D_{6h}, have been reported to be 1.6 and 0.95 eV, respectively[18)]. Using the criterion for symmetry reduction obtained above, we can then conclude that though $C_{18}H_{18}$ does not undergo a symmetry reduction, $C_{30}H_{30}$ should suffer a symmetry reduction $(D_{6h} \rightarrow D_{3h})$ and have alternating bonds in its ground state. The symmetry reduction in $[4n+2]$ annulenes is thus predicted to set in at $n \simeq 5$, which is in agreement with the predictions obtained by Dewar and Gleicher[33)], Binsch and Heilbronner[34)], and Binsch et al.[35)] using a more sophisticated theoretical method.

As for $C_{18}H_{18}$, available experimental facts seem to support the above conclusion. Gouterman and Wagnière[57)] have indicated that the vibronic analysis of the electronic spectrum weighs against the assumption that bond alternation occurs in $C_{18}H_{18}$. Further, the X-ray experiments on $C_{18}H_{18}$[58), 59)] have shown that it has a D_{6h} symmetry (however, the X-ray data do not necessarily exclude the possibility of the existence of rapidly interconverting alternating bonds[13)]). Finally, more conclusive support for the above conclusion is provided by the fact that the lowest excitation energy calculated assuming D_{6h} symmetry is in good agreement with the position of the longest wave-length absorption band $({}^1B_{2u})$ in the electronic spectrum recently discovered by Blattmann et al.[60)].

Finally, we consider the application of the symmetry rule to the prediction of the *geometrical structures of the excited states* of conjugated hydrocarbons. On the basis of the same approximations as we

T. Nakajima

have used in deriving Eq. (11), the force constant for the ith normal nuclear displacement in the mth excited state may be given by

$$f_m^i = k - 2 \sum_{n \neq m} \frac{\left| \left\langle \psi_n \left| \frac{\partial H_\pi}{\partial Q_i} \right| \psi_m \right\rangle \right|^2}{E_n - E_m}. \tag{17}$$

In order to seek the most soft nuclear deformation in an excited state, the approximation is again made of replacing the sum over excited states in Eq. (17) by a dominant term corresponding to the next higher excited state. Now, the transition density ρ_{nm} between the nth excited state corresponding to the orbital jump $\phi_i \to \phi_k$ and the mth excited state corresponding to $\phi_j \to \phi_l$ is $\phi_k \phi_l$ if $i=j$, and $-\phi_i \phi_j$ if

Table 2. Energies and symmetries of the second excited states of nonalternant hydrocarbons

Molecule[a]	$E_2 - E_1$ (eV)	Symmetry	Molecule[a]	$E_2 - E_1$ (eV)	Symmetry
I	3.25	B_{1u}	XV	1.37	A_1
II	2.15	B_{3g}	XVI	0.99	B_{2u}
III	1.54	B_{1u}	XVII	0.17	B_2
IV	1.48	B_2	XVIII	0.98	B_{2u}
V	2.27	B_2	XIX	1.74	A_1
VI	0.79	B_u	XX	1.35	A_1
VII	2.41	B_{1u}	XXI	0.05	B_{2u}
VIII	1.75	B_{2u}	XXII	0.06	B_2
IX	1.20	B_{1u}	XXIII	0.06	B_{2u}
X	1.06	B_2	I^{-2}	0.66	B_{1u}
XI	1.48	A_1	III^{-2}	1.00	B_{1u}
XII	1.51	A_1	IV^{-2}	0.05	A_1
XIII	1.62	A_1	VII^{-2}	0.44	B_{1u}
XIV	1.18	A_1			

[a] See Fig. 5.

$k=l$ (vanishing otherwise). Therefore, the square of the matrix element $|\langle \psi_n | \partial H_\pi / \partial Q_i | \psi_m \rangle|^2$ in Eq. (17) would be about one half of that of $|\langle \psi_n | \partial H_\pi / \partial Q_i | \psi_0 \rangle|^2$ in Eq. (11), in so far as the transition density is assumed not to vary significantly, depending on the molecular orbitals concerned. Thus, if the energy gap $E_{m+1} - E_m$ between an excited state and the next higher one is smaller than ca. 0.6 eV, the force constant for a certain normal nuclear displacement in the excited state of interest is expected to be negative.

In Table 2 are listed the energies of the second excited states (measured from the first excited states) corresponding to the first-order equilibrium nuclear configurations of the ground states for nonalternant hydrocarbons and some of their dianions. Use of these energy values and the corresponding wavefunctions in Eq. (17) is approximative. In a strict sense, the energy values and corresponding wavefunctions to be used in Eq. (17) should be those corresponding to the first-order equilibrium nuclear configuration of the mth excited state as given by minimizing its energy while maintaining the highest molecular symmetry.

Inspection of Table 2 reveals that all those molecules that suffer a molecular-symmetry reduction in the ground state possess $(E_2 - E_1)$ values considerably larger than the critical value, so that they should have a fully-symmetrical nuclear configuration in their first excited states. On the other hand, there are cases where a molecule has an $(E_1 - E_0)$ value significantly higher than the critical value, but has a relatively smaller $(E_2 - E_1)$ value. The $(E_2 - E_1)$ value of the pentalene dianion (I^{-2}) is of the same order of magnitude as the critical value; and those for the peri-condensed nonalternant hydrocarbon, XVII, the fulvalenes, XXI, XXII and XXIII, and the dianions, IV^{-2} and VII^{-2}, are significantly smaller than the critical value ($\simeq 0.6$ eV).

The orbital arrangement for pentalene shown in Fig. 2 serves to indicate how close the second excited state is to the first excited state when two more electrons are placed in the nonbonding orbital to form the dianion. The very small $(E_2 - E_1)$ values for fulvalene and heptafulvalene are realized from the orbital arrangements shown in Fig. 4: in both molecules the two lowest excited states ($^1B_{3g}$ and $^1B_{2u}$) have the same energy in the Hückel picture.

The symmetry of the most soft distortion in the lowest excited state is given by the direct product of the symmetry of the first excited state (shown in Table 1) and that of the second excited state (shown in Table 2). These symmetries are $b_{3g}(R_x)$ for I^{-2} and VII^{-2}, $b_2(y)$ for XVII and IV^{-2}, $b_{1u}(z)$ for XXI and XXIII, and $a_1(z)$ for XXII. The symmetries of the lowest excited states are then predicted to be C_{2h}, C_s, C_{2v} and C_{2v}, respectively. It should be noted that despite the strong vibronic coupling with the second excited state, the first excited state of sesquifulvalene (XXII) does not undergo a symmetry reduction.

Using the same method as described in II.B, Binsch and Heilbronner[37] have examined the second-order bond distortion in the lowest excited states of nonalternant hydrocarbons (I, IV – VII, X, XI, XIII – XV and XVII), and have shown that, of the molecules examined, only VI and XVII suffer a molecular-symmetry reduction in the lowest

23

T. Nakajima

excited state. This is in qualitative agreement with the above results (note that the $(E_2 - E_1)$ value for VI is higher than the critical value, but still very close to it). Futher, partial support for the above results will be provided later by the geometrical structures of lowest excited states of molecules VII, XI, XXI, XXII and XXIII calculated using the dynamic theory.

III. The Dynamic Theory of Bond Distortion

A. Method of Calculation

The method of calculation used for obtaining the energetically most stable set of C—C bond distances in a conjugate molecule is the SCF formalism of the Pariser-Parr-Pople semiempirical MO method[61), 62)], taken together with the variable bond-length technique[63)]. The C—C bond distances and, correspondingly, the resonance and Coulomb repulsion integrals are allowed to vary with bond orders at each iteration until self-consistency is achieved. Bond lengths are correlated with bond orders by the aid of the following formula[17)]

$$r_{\mu\nu}(\text{Å}) = 1.520 - 0.186\, p_{\mu\nu}. \tag{18}$$

The Coulomb repulsion integrals are evaluated using the Mataga-Nishimoto formula[64)]. The resonance integral is assumed to be of exponential form $\beta = B e^{-ar}$, the value of exponent a being taken as $1.7\,\text{Å}^{-1}$ [65)].

As the starting geometries for iterative calculation, we take all the possible structures in which bond lengths are distorted so that the set of displacement vectors may form a basis of an irreducible representation of the full symmetry group of a molecule. For example, with pentalene (I), there are 3, 2, 2 and 2 distinct bond distortions belonging respectively to a_g, b_{3g}, b_{2u} and b_{1u} representations of point group D_{2h}. Further, if a certain distorted structure and its countertype in which the bond-length variation is reversed are not equivalent (e. g., the two Kekulé-type structures in IV), these two structures should be differentiated as a starting geometry.

When self-consistency is achieved at two different geometries, one possessing the full symmetry and the other possessing a lower symmetry, the latter should, in principle, be more stable than the former. In such a case, we define the stabilization energy as the difference in total energy between the two structures. The total energy is calculated using Eqs. (3) and (4). The value of the force constant k adopted for use in calculating the σ-core energy using Eq. (4) is $714\ \text{kcal mole}^{-1}\,\text{Å}^{-2}$ [15)].

In order to discuss the geometrical structures of electronically excited states, the same procedure as described above is used, except for the use of a different value 3.3 Å^{-1} for exponent a in the exponential form of the resonance integral[66]. This value of a was determined so that the predicted fluorescence energy from the lowest singlet excited state ($^1B_{2u}$) in benzene may fit the experimental value.

B. Results and Discussion

The symmetry groups and bond lengths corresponding to the most stable nuclear configurations for nonalternant hydrocarbons and some of their dianions are shown in Fig. 5.

In cata-condensed nonalternant hydrocarbons (I − XIII), except IV, VI, X and XI, two different self-consistent nuclear arrangements, one belonging to the full symmetry group of a molecule and the other belonging to a reduced symmetry group, were obtained. In pentalene, for example, the starting bond distortions, belonging to a_g, b_{2u} and b_{1u} representations, all converge into the unique self-consistent set of bond lengths belonging to point group D_{2h}, and those belonging to b_{3g} converge into the set of bond lengths belonging to point group C_{2h}. The stabilization energies which favor the lower-symmetry nuclear arrangements for molecules I, II, III, V, VII, VIII, IX, XII and XIII are calculated to be 8.4, 0.6, 2.4, 7.2, 12.1, 5.3, 5.9, 2.5 and 6.6 kcal mole^{-1}, respectively. All these molecules exhibit in a greater or lesser extent a marked bond alternation, the degree of which may be estimated from the magnitude of stabilization energy; in the fully symmetrical equilibrium structures of these molecules, bond lengths of the peripheral C—C bonds are nearly equalized.

In Fig. 6 the stabilization energies (ΔE_s) are plotted against the lowest excitation energies ($E_1 - E_0$), as calculated assuming the fully-symmetrical structures. It is seen that there is a good correlation between these quantities. This indicates that the seemingly drastic approximation that only the vibronic coupling between the lowest excited state and the ground state plays a dominant role in estimating the force constant for the most soft nuclear deformation in the ground state of a conjugated molecule is actually justified not only qualitatively but also quantitatively.

On the other hand, in cata-condensed nonalternant hydrocarbons IV, VI, X and XI, peri-condensed nonalternant hydrocarbons XIV − XVIII, fulvenes XIX and XX, and fulvalenes XXI − XXIII, self-consistency was achieved only for the fully-symmetrical nuclear arrangement. All these molecule, except azulene (XI), also show in a greater or lesser degree a pronounced double-bond fixation.

T. Nakajima

Fig. 5 (p. 26–29). Predicted and experimental (in italics) bond lengths (in Å) of non-alternant hydrocarbons and choice of axes (Experimental data: for XI, Robertson, J. M., Shearer, M. M., Sim, G. A., Watson, D. G.: Acta Crist. *15*, 1 (1962); for XIV (a dimethyl derivative) see Reference 72); for the upper values of XV (a dimethylphenyl derivative) see Reference 73); for the lower values of XV (a tetramethyl derivative) see Reference 74). For VI the *z* axis is chosen perpendicular to the molecular plane)

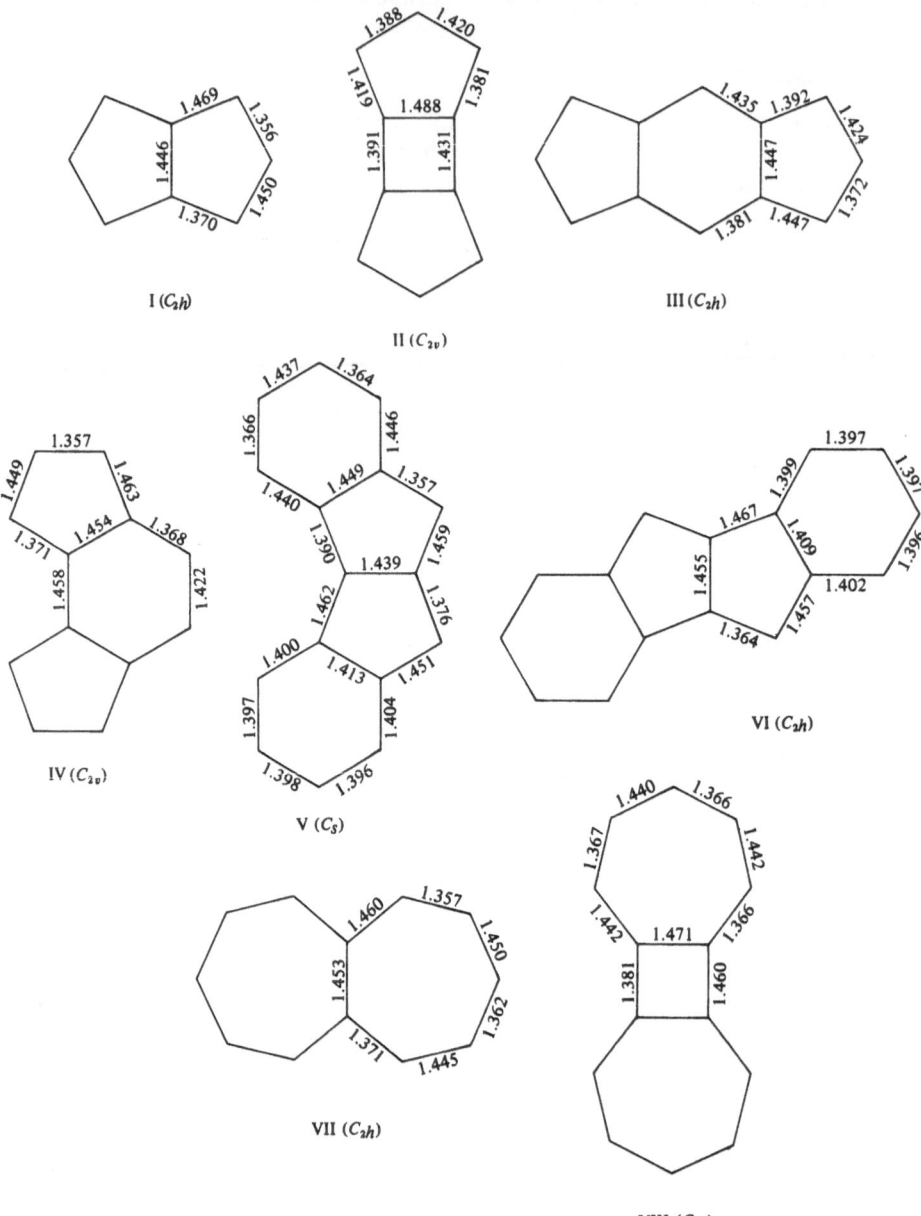

Fig. 5 (continued)

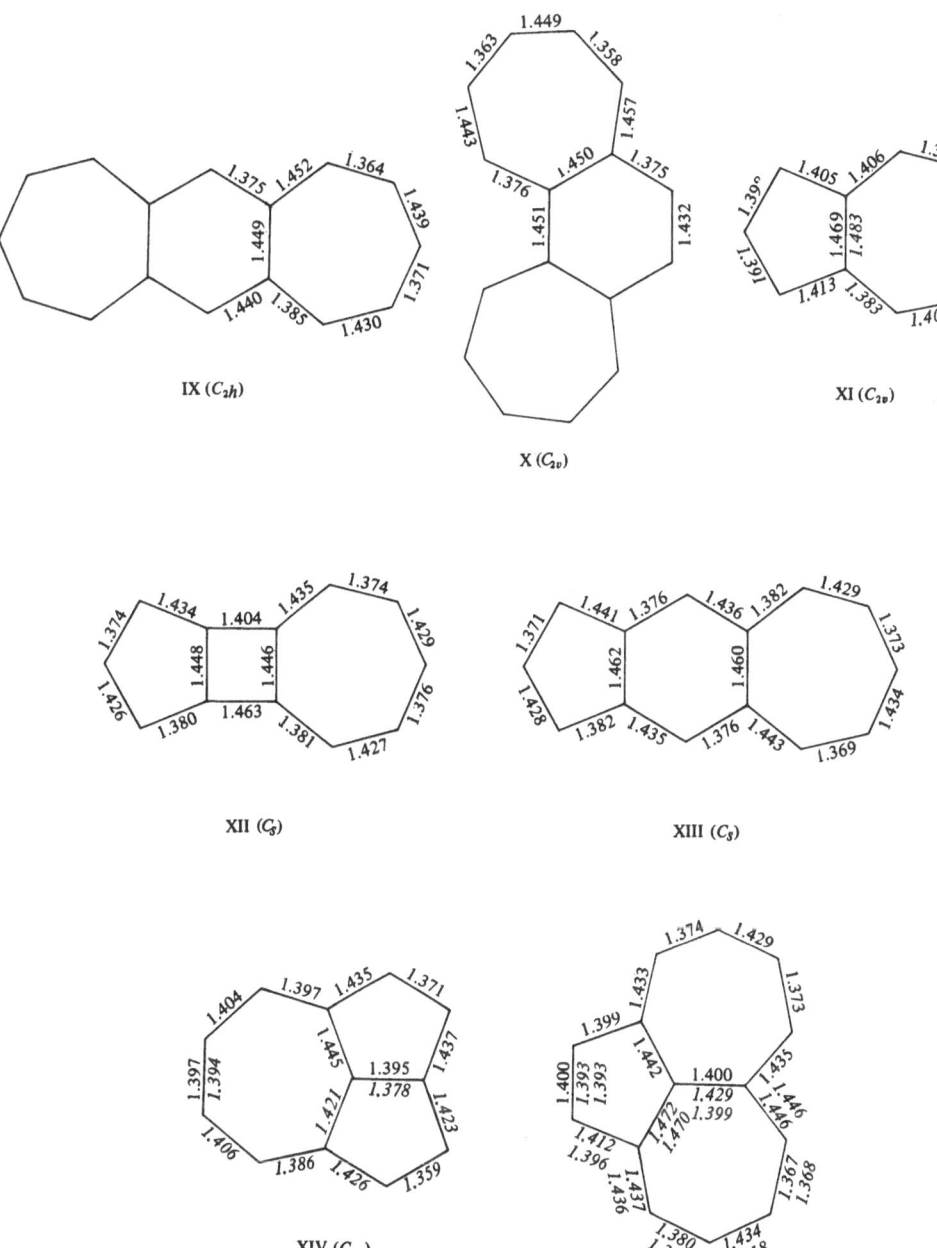

IX (C_{2h})

X (C_{2v})

XI (C_{2v})

XII (C_{s})

XIII (C_{s})

XIV (C_{2v})

XV (C_{2v})

27

Fig. 5 (continued)

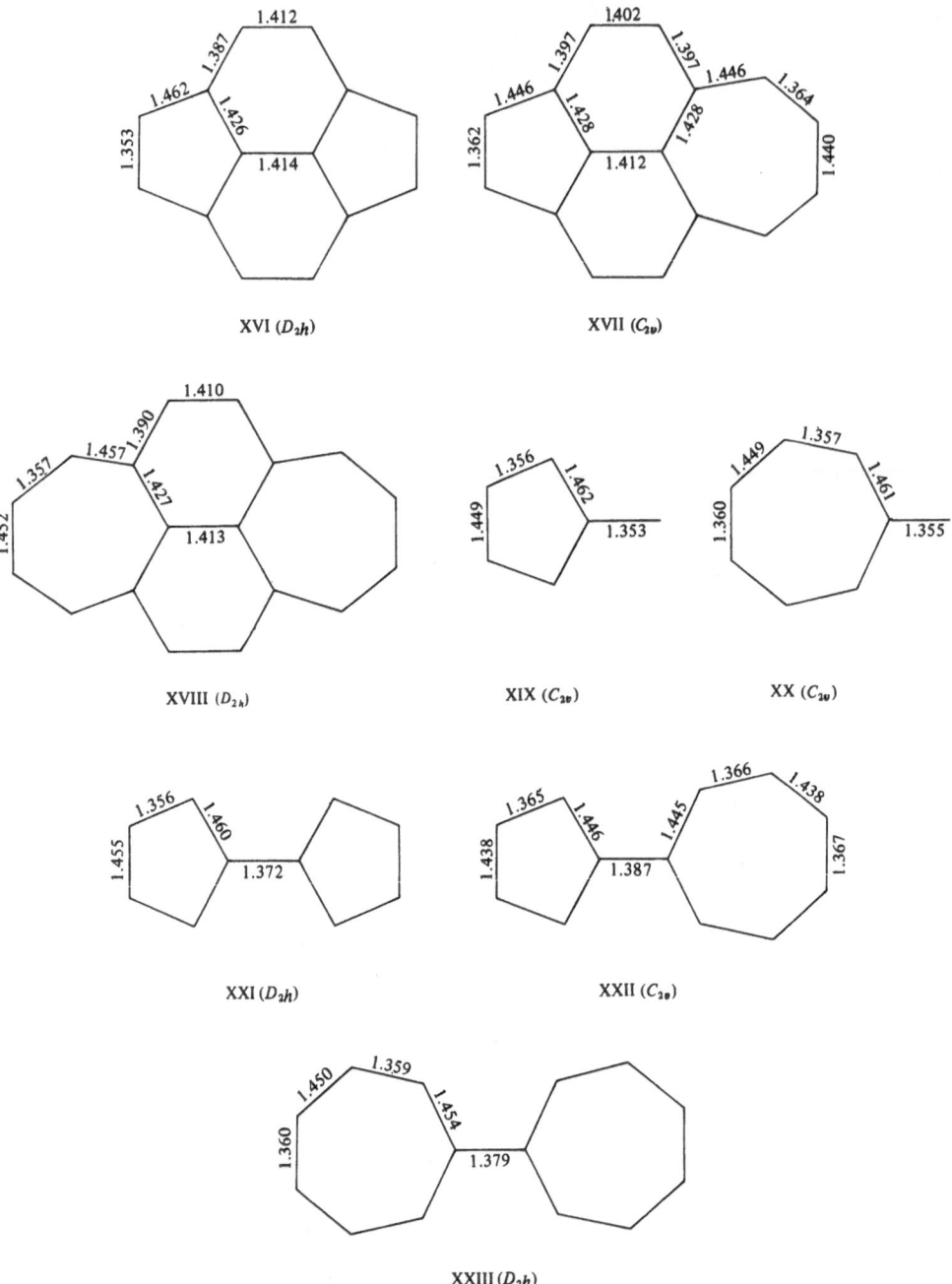

XVI (D_{2h})

XVII (C_{2v})

XVIII (D_{2h})

XIX (C_{2v})

XX (C_{2v})

XXI (D_{2h})

XXII (C_{2v})

XXIII (D_{2h})

Fig. 5 (continued)

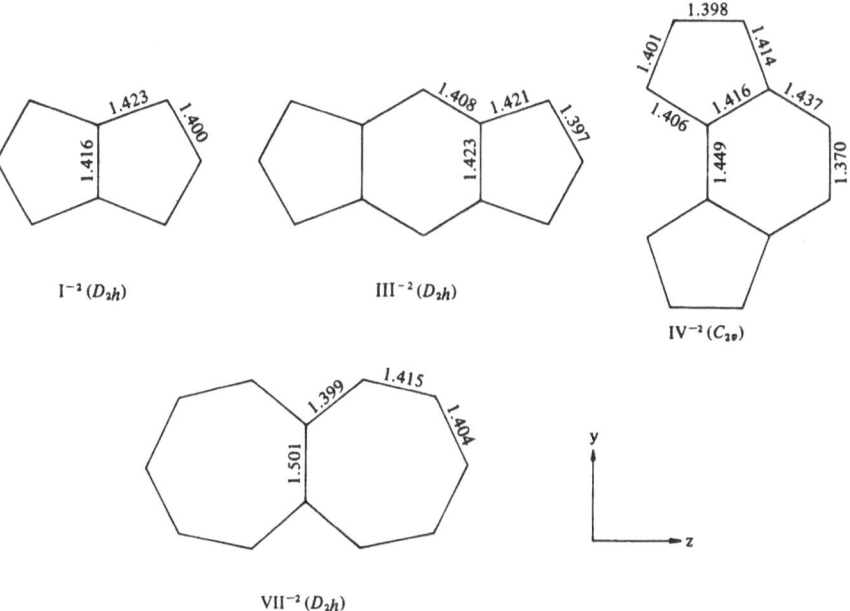

$I^{-2} (D_2h)$

$III^{-2} (D_2h)$

$IV^{-2} (C_{2v})$

$VII^{-2} (D_2h)$

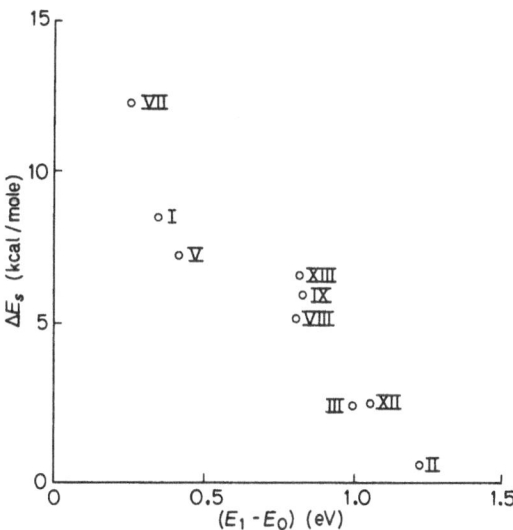

Fig. 6. Correlation of $(E_1 - E_0)$ with ΔE_s

Of the cata-condensed nonalternant hydrocarbons undergoing a pseudo Jahn-Teller distortion, pentalene (I) and heptalene (VII), having the largest ΔE_s value, are predicted to possess a strong bond alternation. This confirms the results of the previous theoretical investigations[14]−[17], [19] and agrees with the available experimental facts[6], [7].

The s-indacene (III) synthesized by Hafner's group[67] and its 7-membered analogue (IX) are predicted to assume a skewed structure with a moderate double-bond fixation. As for (III), this conclusion supports the previous theoretical works[68] and is in accordance with the experimental information.

Of particular interest is dibenzopentalene (V), which suffers a strongly-antisymmetrical bond distortion. In one of the 6-membered rings of this molecule bond lengths are nearly equal as in benzene itself, whereas in the other 6-membered ring as well as in the pentalene segment strong double-bond fixations exist. In this respect the static theory is insufficient: the transition density of ρ_{01} of this molecule (Fig. 1) indicates that strong bond fixations are localized in the pentalene segment, but in both the 6-membered rings only a slight degree of bond fixation exists.

In the isomeric dibenzopentalene (VI), which undergoes only a first-order bond fixation, the two 6-membered rings are equivalent and in both the rings bond lengths are almost equalized. Strong bond fixations are localized in the pentalene-like region. It is therefore expected that this molecule would undergo addition reactions in this region in accordance with the experimental facts[69], [70].

Other molecules worth mentioning are peri-condensed nonalternant hydrocarbons, XIV − XVIII, which undergo a first-order bond distortion. As for XIV and XV, it is noted that bond lengths of the peripheral bonds of XIV belonging to the 7-membered ring and those of XV belonging to the 5-membered ring are all 1.4 Å, while in both the molecules a strong bond fixation exists in the other region of the molecular periphery[71]. Dimethyl derivatives of XIV and XV have been prepared by Hafner and Schneider[8], and the predicted bond lengths are in good agreement with the X-ray data[72]−[74].

For molecules XVI − XVIII it is predicted that bond lengths of the naphthalene core are almost the same as those of the free naphthalene molecule, and in the other region of the molecule marked bond fixations are observed[75]. The most stable ground-state geometry of XVI corresponds to the "aromatic model" proposed by Lo and Whitehead[76]. Molecules XVI and XVII have been synthesized by Trost and Bright[77] and Boekelheide and Vick[78], respectively. The present results are in good qualitative agreement with the available experimental information.

It is added that Tajiri et al.[79] have recently examined the geometrical structures of fundamental nonalternant hydrocarbons I, VII, XI, XIX, XX, and XXI—XXIII using the CNDO/2 SCF MO method, in which the assumption of $\sigma - \pi$ separation is lifted by taking into account all the valence electrons[80],[81]. Theoretical C—C bond lengths obtained by minimizing the total energy are in good agreement with those calculated using the dynamic theory.

The bond lengths for the dianions of I, III, IV and VII shown in Fig. 5 indicate that these dianions all have a fully-symmetrical nuclear arrangement and that the bond lengths of the peripheral bonds of these anions are fairly equivalent. This means, in agreement with the results obtained using the static theory, that addition of two more electrons to molecules I, III and VII to form their dianions results in a complete disappearance of the bond alternations existing in these molecules. Dianions of I[82],[83], III[84], and IV[85],[86] have been prepared and are known to be stable species.

Next, we discuss the symmetries, bond lengths and spin densities for the anion and cation radicals of fulvalene (XXI) and heptafulvalene (XXIII) using the dynamic theory. We use the semiempirical open-shell SCF MO formalism[51],[52], in conjunction with the variable bond-length technique.

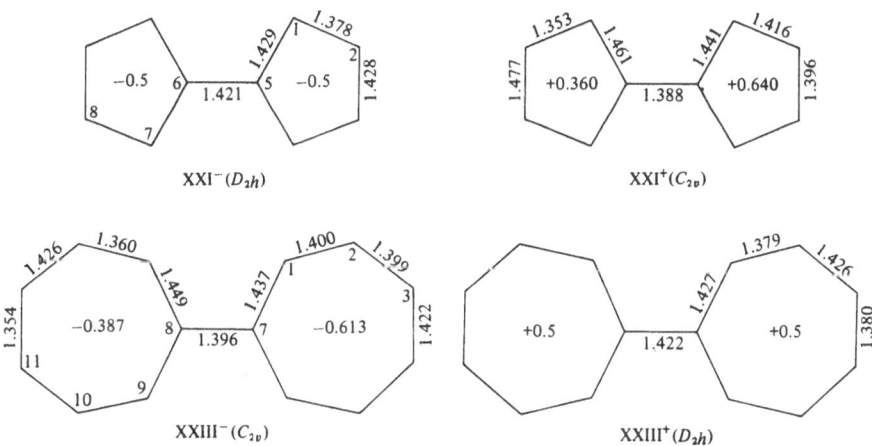

Fig. 7. Bond lengths (in Å), symmetries, ring charge densities and numberings

In the anion radical of XXIII, the starting distorted structures belonging to the a_g, b_{3g} and b_{2u} irreducible representations of point group D_{2h} all converge into the unique set of bond lengths corresponding to

D_{2h}, and the distorted structures belonging to b_{1u} converge into the other set of bond lengths corresponding to point group C_{2v}. The stabilization energy which favors the C_{2v} structure is calculated to be 9.4 kcal mole^{-1}.

The calculated bond lengths for the C_{2v} structure of the anion radical of heptafulvalene shown in Fig. 7 indicate that in one of the ring there exists a significant bond fixation to the same extent as that in the neutral heptafulvalene, while in the other ring bond lengths are nearly equalized. The calculated spin densities, presented in Table 3, indicate that the unpaired spin is localized essentially on the latter ring.

The cation radical heptafulvalene, on the other hand, undergoes no symmetry reduction. Both the rings show a moderate double-bond fixation (Fig. 7), and the unpaired spin is delocalized throughout the molecule (Table 3).

Table 3. Spin densities (ρ) and calculated and experimental hyperfine splittings ($|a^H|$ in G)

Radical	Atom[a]	ρ	$|a^H|_{cal}$	$|a^H|_{exp}$
XXI$^-$ (D_{2h})	1	0.065	1.69	
	2	0.097	2.52	
	5	0.186		
XXI$^+$ (C_{2v})	1	0.379	9.85	
	2	0.120	3.12	
	5	0		
	6	0		
	7	0.000		
	8	0.000		
XXIII$^-$ (C_{2v})	1	0.275	715	8.22
	2	0.040	1.04	0.3
	3	0.189	4.91	5.02
	7	0		
	8	0		
	9	0.000		
	10	0.000		
	11	0.000		
XXIII$^+$ (D_{2h})	1	0.043	1.09	0.075
	2	0.082	2.13	2.90
	3	0.052	1.53	1.72
	7	0.133		

[a] See Fig. 7.

In Table 3 are listed the proton hyperfine splittings (a^H) for the anion and cation radicals of heptafulvalene, calculated using McConnell's relationship[87)] with $|Q| = 26$ G, together with the experimental values. Theoretical values are in fairly good agreement with experimental data.

In the anion and cation radicals of fulvalene (XXI) the situation turns out to be quite reversed. Removal of an electron from the neutral molecule to produce the cation radical results in a symmetry reduction $(D_{2h} \rightarrow C_{2v})$, the stabilization energy being calculated to be 17.8 kcal mole^{-1}. On the other hand, addition of an electron to form the anion radical leaves the molecular symmetry unaffected.

Inspection of Fig. 7 reveals that in the cation radical of XXI a marked bond fixation exists in one of the rings, while bond lengths are nearly equalized in the other ring, on which the unpaired spin is delocalized (Table 3). On the other hand, in the anion radical of XXI, there is a moderate bond fixation in both rings, and the unpaired spin is delocalized over the entire molecule.

Finally, electronically excited states are considered in the framework of the dynamic theory. In Fig. 8 are shown calculated bond lengths in the lowest excited states of heptalene (VII), azulene (XI) and the fulvalenes (XXI – XXIII).

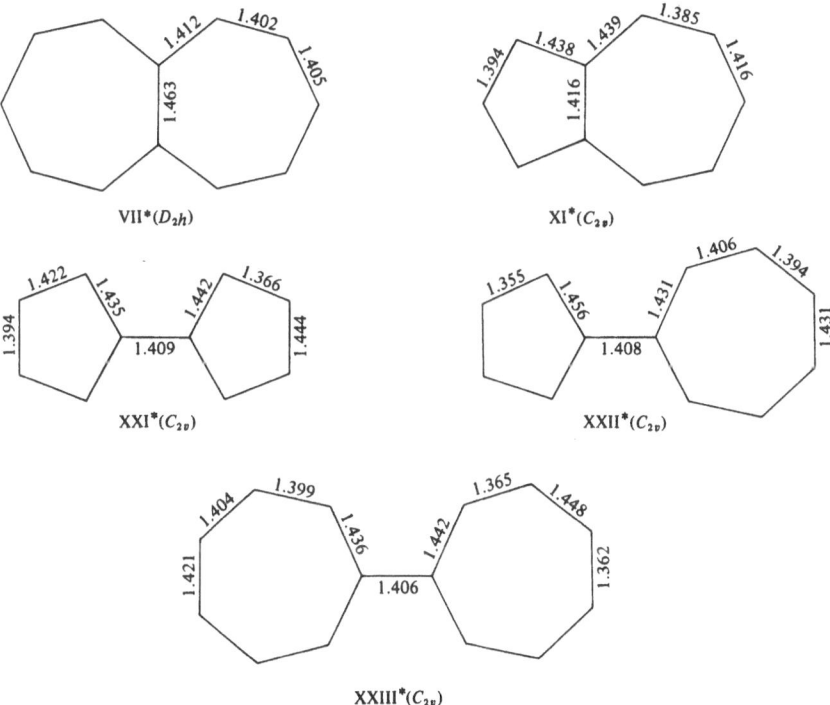

Fig. 8. Bond lengths (in Å) and symmetries of the lowest excited states

33

In the lowest excited state of heptalene all the distorted structures give rise to a unique set of bond lengths belonging to point group D_{2h}, the result supporting the prediction obtained on the basis of the symmetry rule. In the lowest excited state bond lengths of the peripheral bonds are highly equalized. The marked difference in geometrical structure between the ground state and the lowest excited state may provide the main reason for the appearance of a long absorption tail throughout the visible region in the electronic spectrum of heptalene[6].

The lowest excited state of azulene is predicted to possess C_{2v} symmetry, which is in agreement with the result obtained using the symmetry rule. A recent vibrational analysis of the longest wave-length absorption band in the electronic spectrum of azulene indicates that the lowest-excited state would possess C_{2v} symmetry[88].

Fulvalene and heptafulvalene are predicted, in agreement with the result obtained using the symmetry rule, to suffer a symmetry reduction $D_{2h} \rightarrow C_{2v}$ in their lowest excited states. The longest wave-length electronic absorption bands of these molecules are expected to be relatively broad. This seems to be what is observed[3, 4]. On the other hand, the lowest excited state of sesquifulvalene is predicted not to undergo symmetry reduction, which again supports the prediction based on the symmetry rule.

IV. Magnetic Susceptibilities

The anisotropy of the magnetic susceptibility of a cyclic conjugated system, attributable to induced ring currents in its π-electron network, is one of the important quantities indicative of π-electron delocalization. The method used for the calculation of the magnetic susceptibilities of nonalternant hydrocarbons is the London-Hoarau method[89] taken together with the Wheland-Mann SCF technique[13]. The resonance integral is assumed again to be of exponential form $\beta = Be^{-ar}$, but the value of a now used is 4.4 Å^{-1} [90]. The Wheland-Mann SCF MO method with this value of a reproduces well the bond orders and electron densities of nonalternant hydrocarbons obtained using the dynamic theory.

In Table 4 theoretical magnetic anisotropies ΔK for selected molecules, calculated assuming the bond lengths shown in Fig. 5, are listed (in units of $\Delta K_{\text{benzene}}$) and compared with the experimental exaltations reported by Dauben et al.[91]. Theoretical values are in good agreement with experimental data, except for aceheptylene (XV), for which experimental exaltation has been reported to be zero (as to the theoretical values for XIV and XV, see also 92))!

The magnetic susceptibilities for molecules I, III, VII, XVI and XVIII were predicted to be negative (in units of that of benzene), that is, to be paramagnetic. Mayo et al.[93] and Pullman et al.[9), 94] first pointed out that in certain nonalternant hydrocarbons, such as pentalene and heptalene, the contribution due to π-electron delocalization to the magnetic susceptibility turns out to be paramagnetic instead of diamagnetic. Such an anomalous magnetic susceptibility is currently interpreted as being due to an induced or quenched paramagnetism[95)-98]. According to Van Vleck[99], the magnetic susceptibility of a molecule is given by the coefficient of H^2 in the perturbational expansion of the total energy in terms of the external magnetic field H; and the molar magnetic susceptibility is expressed by

$$\chi_z = -\frac{N e^2}{4 m c^2} \langle \psi_0 | \sum_i (x_i^2 + y_i^2) | \psi_0 \rangle + \frac{N e^2}{2 m^2 c^2} \sum_{n \neq 0} \frac{|\langle \psi_n | M_z | \psi_0 \rangle|^2}{E_n - E_0} \qquad (19)$$

where (1) the z axis is chosen perpendicular to the molecular plane, (2) x_i and y_i are the distances of electron i to the origin, and (3) M_z is the operator for the z component of the orbital angular momentum of the electrons. The first term of Eq. (19) represents the pure diamagnetism, while the second term is the Van Vleck or second-order temperature-independent paramagnetic term. We notice at once the resemblance between Eq. (19) and the term in braces of Eq. (11), and realize that, when there are very low-lying excited states, the second term may be large enough to overcome the first term and bring about a net paramagnetic susceptibility. Indeed molecules I, III and VII, which show a paramagnetic susceptibility, possess a very low-lying lowest excited state *(vide infra)*, in which the symmetry is such that $\langle \psi_1 | M_z | \psi_0 \rangle \neq 0$ [However, we cannot expect a reasonable correlation between the $(E_1 - E_0)$ value and the paramagnetic susceptibility because the first term (the diamagnetic term) cannot be considered to be constant even approximately from molecule to molecule: molecules XVI and XVIII, though their lowest excitation energies are not so small, do exhibit a paramagnetic susceptibility.]

For molecules I, III and VII, Table 4 also gives theoretical values corresponding to the fully-symmetrical equilibrium structures. It will be seen that double-bond fixation brings about a considerable decrease in paramagnetic susceptibility. Partial evidence for such reduced magnetic susceptibilities for III[67] and XVI[77] is provided by their proton NMR spectra, which show proton signals in the olefinic region.

It is concluded that the magnetic susceptibility of a conjugated molecule is a very sensitive indicator of bond distortion; and, in the

Table 4. Calculated magnetic susceptibilities (ΔK) and experimental exaltation (Λ)

Molecule	$\Delta K/\Delta K_{benzene}$	$\Lambda/\Lambda_{benzene}$
I (D_{2h})	−1.35	
I (C_{2h})	−0.41	
III (D_{2h})	−0.88	
III (C_{2h})	−0.38	
VI (C_{2h})	1.10	1.00[a], 0.87[b],[c]
VII (D_{2h})	−4.41	
VII (C_{2h})	−0.61	−0.45[a]
XI (C_{2v})	2.04	2.16[a]
XIV (C_{2v})	2.14	2.18[a],[d]
XV (C_{2v})	1.72	0.0[a],[c]
XVI (D_{2h})	−0.40	
XVII (C_{2v})	3.92	3.87[a]
XVIII (D_{2h})	−0.34	
XIX (C_{2v})	0.10	0.08[a]
XX (C_{2v})	0.16	
XXI (D_{2h})	0.075	
XXIII (D_{2h})	0.18	0.15[a]

[a] Reference 91).
[b] Reference 89).
[c] The value for a dimethyl derivative.
[d] The value for a dimethylphenyl derivative.

nonalternant hydrocarbons examined, the magnetically induced ring current is very much impeded as compared with that expected on the basis of Pauling's free-electron model[13], except for azulene.

V. Electronic Spectra

The method used for calculating excitation energies is the Pariser-Parr-Pople SCF CI MO method with the same parameterizations as those used in determining the ground-state geometries. Configuration mixing of all the singly excited states is included.

In Table 5 are summarized theoretical lower-excitation energies and corresponding intensities for nonalternant hydrocarbons for which experimental data are available. Theoretical values are in general agreement with experimental ones. The predicted excitation energies for heptalene (VII) are rather small as compared with experimental data. This may arise as a consequence of nonplanarity due to the steric repulsions between ortho-hydrogen atoms.

It is added that for molecules I, III and VII the excitation energies, particularly the lowest ones, calculated assuming the full symmetries (D_{2h}) (cf. Table 1), turn out to be considerably smaller than those calculated assuming the reduced symmetries (C_{2h}) and experimental values. Recent calculations on the electronic spectra of pentalene and heptalene by Fernández-Alonso and Palou[100] do not agree with this conclusion.

Table 5. Singlet transitions of nonalternant hydrocarbons

Molecule	Transition symmetry[a]	Theoretical		Experimental $E(eV)$
		$E(eV)$	$f(cgs)$	
I (C_{2h})	A_g	1.59	Forb.	1.72 (log ε = 1.95, tailing)[b], 2.00[c]
	B_u	3.78	0.30	3.27 (3.99)[b], 3.70[c]
	B_u	4.75	0.18	4.00 (4.52)[b], 4.35[c]
III (C_{2h})	A_g	1.39	Forb.	1.77 (log ε = 2.59, tailing)[d]
	B_u	2.62	0.59	2.42 (4.63)
	B_u	3.78	0.21	3.61 (4.81)
	A_g	3.88	Forb. ⎫	
	A_g	4.34	Forb. ⎬	4.06 (shoulder), 4.32 (4.68)
	B_u	4.81	1.93 ⎭	
VI (C_{2h})	A_g	2.54	Forb. ⎫	
	B_u	3.32	0.64 ⎬	2.59 (log ε = 4.28)[e]
	A_g	4.23	Forb. ⎫	
	B_u	4.28	0.12 ⎬	4.57 (4.64)
	B_u	4.91	1.29 ⎭	
VII (C_{2h})	A_g	1.57	Forb.	Tail[f]
	B_u	3.02	0.31	3.52 (f = 0.15)
	B_u	3.82	0.17	4.84
XI (C_{2v})	B_2	2.05[g]	0.025[g]	1.96[h], 2.14[i] (f = 0.009)[h],[i]
	A_1	3.53	0.005	3.66[h], 3.50[i] (0.08)[h],[i]
	B_2	4.41	0.13	4.48[h]
	A_1	4.78	1.88	4.52[h],[i] (1.10)[h],[i]
	B_2	5.68	0.41	5.24[h],[i] (0.38)[h],[i]

Table 5 (continued)

Molecule	Transition symmetry[a]	Theoretical		Experimental E(eV)
		E(eV)	f(cgs)	
XIV (C_{2v})	B_2	1.73[j]	0.003[j]	1.84[k]
	A_1	2.91	0.001	2.59
	A_1	3.64	0.073	3.37
	B_2	3.86	0.11	3.79
	B_2	4.79	0.39	4.63
XV (C_{2v})	B_2	1.52[j]	0.014[j]	1.55[k]
	A_1	2.89	0.009	2.92
	B_2	3.47	0.30	3.33
	A_1	3.82	0.14 ⎫	
	B_2	4.48	0.019 ⎬	3.96 ~ 4.77
	A_1	4.57	0.53 ⎭	
	B_2	4.95	1.44	4.94
XVI (D_{2h})	B_{3g}	2.24	Forb.	Tail[l]
	B_{2u}	3.23	0.048	2.91
	B_{1u}	3.75	0.33	3.65
	B_{2u}	4.09	0.18	3.81
	A_g	4.62	Forb.	5.23
	B_{2u}	5.41	0.008	
XVII (C_{2v})	A_1	2.98	0.28 ⎫	
	B_2	3.15	0.026 ⎪	
	B_2	3.32	0.066 ⎬	2.20 ~ 3.54[m]
	B_2	3.87	0.025 ⎭	
	A_1	4.50	0.96 ⎫	
	A_1	4.67	0.16 ⎭	4.14
	B_2	5.09	0.87	4.98
XIX (C_{2v})	B_2	3.32[g]	0.035[g]	3.32[n], 3.42[o] ($f = 0.012$)[n]
	A_1	5.06	0.63	5.12[n],[o] (0.32)[n]
XX (C_{2v})	B_2	2.97[g]	0.047[g]	2.91 ($f = 0.02$)[p]
	A_1	4.32	0.48	4.44(0.3)
	B_2	6.05	0.096	5.83
XXI (D_{2h})	B_{3g}	2.39	Forb.	Tail[q]
	B_{2u}	2.44	0.016	2.98 ($\log \varepsilon = 2.41$)
	B_{1u}	3.87	1.18	3.95(4.67)
XXII (C_{2v})	B_2	2.72	0.008	Tail[r]
	B_2	2.78	0.034	Shoulder
	A_1	3.15	1.11	3.05 ($\log \varepsilon = 4.38$)
	A_1	3.67	0.002	
	A_1	5.12	0.28	5.53(4.20)

Table 5 (continued)

Molecule	Transition symmetry[a]	Theoretical		Experimental E(eV)
		E(eV)	f(cgs)	
XXIII (D_{2h})	B_{3g}	2.14	Forb. ⎱	Tail[q]
	B_{2u}	2.20	0.015 ⎰	
	B_{1u}	3.14	1.31	3.42 (log $\varepsilon = 4.35$)
	A_g	4.93	Forb. ⎫	
	A_g	5.03	Forb. ⎬ 5.30 (4.35)
	B_{3g}	5.28	Forb. ⎪	
	B_{2u}	5.39	0.26 ⎭	

[a] For molecules which belong to the point group C_{2h}, the x axis is taken to be perpendicular to the molecular plane.

[b] The spectrum of hexaphenylpentalene: See LeGoff, E.: J. Am. Chem. Soc. *84*, 3975 (1962).

[c] The spectrum of a methyl derivative; Reference 7).

[d] The spectrum of a hexacarbometoxydihydroxy derivative; LeGoff, E., LaCount, R.B.: Tetrahedron Letters *964*, 1161.

[e] The spectrum of a dichlor derivative; Reference 69).

[f] Reference 6).

[g] Reference 65).

[h] Heilbronner, E.: Nonbenzenoid aromatic compounds (ed. D. Ginsburg). New York: Interscience Publishers 1959.

[i] Plattner, Pl. A.. Heilbronner, E.: Helv. Chim. Acta *30*, 910 (1947); *31*, 804 (1948).

[j] See Reference 71).

[k] The spectrum of a dimethyl derivative; Reference 8) and Hafner, K.: Personal communication.

[l] Reference 77) and Trost, B. M.: Personal communication.

[m] Reference 78).

[n] Reference 2).

[o] Schaltegger, H., Neuenschwander M., Meuche, D.: Helv. Chim. Acta *48*, 955 (1965).

[p] Reference 4).

[q] Reference 3).

[r] The spectrum of a p-methoxybenzyl derivative; Reference 5) and Prinzbach, H.: Personal communication.

VI. Conclusion

The results of our calculations based on both the static and dynamic theories show that most of the nonbenzenoid cyclic conjugated systems examined exhibit in a greater or lesser degree a *marked double-bond fixation.* The static theory indicates that even in benzene there exists a hidden tendency to distort into a skewed structure and that such a tendency is actually realized in $[4n+2]$ annulenes larger than a certain critical size. In nonalternant hydrocarbons bond distortion is a rather common phenomenon. Fulvenes, fulvalenes and certain peri-condensed nonalternant hydrocarbons undergo a first-order bond distortion, and most of the cata-condensed nonalternant hydrocarbons suffer a second-order pseudo Jahn-Teller bond distortion.

Finally, a brief comment about the theoretical aromaticity criterion is relevant. Our calculations revealed that in molecules V, VI and XIV — XVIII the carbon skeleton may be divided into two distinguishable parts: one in which bond lengths are nearly equalized and the other in which a strong bond fixation exists, that is, according to the aromaticity criterion based on bond fixation, one which is aromatic and the other polyolefinic. For such molecules the current theoretical aromaticity criteria, such as the magnitude of delocalization energy or that of magnetic ring current, are not useful. These quantities are associated with the properties of a molecule as a whole and cannot reflect the local behaviors of π-electrons.

Acknowledgment. The manuscript of this contribution had been written while the author was residing at the Department of Chemistry, North Dakota State University as a National Science Foundation Senior Foreign Scientist Fellow. The author would like to thank Professor J. M. Sugihara, Professor S. P. Pappas and the staff of the Department of Chemistry, N. D. S. U. for having given him this opportunity and warm hospitality. His thanks are due in particular to Professor S. P. Pappas who kindly took the trouble to read the manuscript. It is also a pleasure to thank Dr. A. Toyota for his kind cooperation in preparing the manuscript.

References

[1] Pfau, A. S., Plattner, P. A.: Helv. Chim. Acta *19*, 858 (1936).
[2] Thiec, J., Wiemann, J.: Bull. Soc. Chim. France *1956*, 177.
[3] Doering, W. v. E., in: Theoretical organic chemistry, p. 35. New York: Academic Press 1959.
[4] — Willey, D. W.: Tetrahedron *11*, 183 (1960).
[5] Prinzbach, H., Rosswog, W.: Angew. Chem. *73*, 543 (1961).
[6] Dauben, H. J., Bertelli, D. J.: J. Am. Chem. Soc. *83*, 4659 (1961).
[7] Bloch, R., Marty, R. A., de Mayo, P.: J. Am. Chem. Soc. *93*, 3071 (1971).
[8] Hafner, K., Schneider, J.: Angew. Chem. *70*, 702 (1958); Liebigs Ann. Chem. *624*, 37 (1959); *672*, 194 (1964).
[9] Pullman, B., Pullman, A.: Les Théories électroniques de la Chimie organique. Paris: Masson 1952.
[10] Hückel, E.: Z. Physik *70*, 204 (1931); *76*, 628 (1932); Z. Elektrochem. *43*, 752 (1937).
[11] Sondheimer, F., Wolovsky, R., Amiel, Y.: J. Am. Chem. Soc. *87*, 3253 (1965).
[12] Jackman, L. M., Sondheimer, F., Amiel, Y., Ben-Efraim, D. A., Gaoni, Y., Wolovsky, R., Bothner-By, A. A.: J. Am. Chem. Soc. *87*, 4307 (1962).
[13] Salem, L.: The molecular orbital theory of conjugated systems. New York: Benjamin 1966.
[14] Boer-Veenendaal, P. C., Vliegenthart, J. A., Boer, D. H. W.: Tetrahedron *18*, 1325 (1962).

[15] Snyder, L. C.: J. Phys. Chem. *66*, 2299 (1962).
[16] Nakajima, T., Yaguchi, Y., Kaeriyama, R., Nemoto, Y.: Bull. Chem. Soc. Japan *37*, 272 (1964).
[17] — Katagiri, S.: Mol. Phys. *7*, 149 (1963).
[18] Longuet-Higgins, H. C., Salem, L.: Proc. Roy. Soc. (London) *A257*, 445 (1960).
[19] Boer-Veenendaal, P. C., Boer, D. H. W.: Mol. Phys. *4*, 33 (1961).
[20] Kuhn, H.: J. Chem. Phys. *16*, 840 (1948); *17*, 1198 (1949); Angew. Chem. *69*, 239 (1957); *71*, 93 (1959).
[21] Lennard-Jones, J. E.: Proc. Roy. Soc. (London) *A158*, 280 (1937).
[22] Coulson, C. A.: Proc. Roy. Soc. (London) *A164*, 383 (1938); *A169*, 413 (1939).
[23] Dewar, M. J. S.: J. Chem. Soc. *1952*, 3546.
[24] Simpson, W. T.: J. Am. Chem. Soc. *73*, 5363 (1951); *77*, 6164 (1955); *78*, 3585 (1956).
[25] Platt, J. R.: J. Chem. Phys. *25*, 80 (1956).
[26] Huzinaga, S., Hashino, T.: Progr. Theoret. Phys. *18*, 649 (1957). See also, Tsuji, M., Huzinaga, S., Hashino, T.: Rev. Mod. Phys. *32*, 425 (1960).
[27] Labhart, H.: J. Chem. Phys. *27*, 957 (1957).
[28] Ooshika, Y.: J. Phys. Soc. Japan *12*, 1238, 1246 (1957).
[29] Longuet-Higgins, H. C., Salem, L.: Proc. Roy. Soc. (London) *A251*, 172 (1959).
[30] Coulson, C. A., Dixon, T. W.: Tetrahedron *17*, 215 (1962).
[31] Förstering, H. D., Huber, W., Kuhn, H.: J. Quant. Chem. *1*, 240 (1967).
[32] Nakajima, T., in: Molecular orbitals in chemistry, physics and biology, p. 457. New York: Academic Press 1964.
[33] Dewar, M. J. S., Gleicher, G. J.: J. Am. Chem. Soc. *87*, 685 (1965).
[34] Binsch, G., Heilbronner, E., in: Structural chemistry and molecular biology, p. 815. San Francisco: Freeman 1968.
[35] — Tamir, I., Hill, R. D.: J. Am. Chem. Soc. *91*, 2446 (1969).
[36] — Heilbronner, E., Murrell, J. N.: Mol. Phys. *4*, 305 (1966).
[37] — — Tetrahedron *24*, 1215 (1968).
[38] — Tamir, I.: J. Am. Chem. Soc. *69*, 2450 (1969).
[39] Nakajima, T., Toyota, A.: Chem. Phys. Lett. *3*, 272 (1969).
[40] — — Yamaguchi, H., in: Aromaticity, pseudoaromaticity, antiaromaticity. New York: Academic Press 1971.
[41] Nakajima, T.: Pure and Appl. Chem. *28*, 219 (1971).
[42] — Toyota, A., Fujii, S.: Bull. Chem. Soc. Japan *45*, 1022 (1972).
[43] Hurley, A. C., in: Molecular orbitals in chemistry, physics and biology, p. 61. New York: Academic Press 1964.
[44] Pearson, R. G.: J. Am. Chem. Soc. *91*, 1252, 4947 (1969).
[45] — J. Chem. Phys. *52*, 2167 (1970).
[46] Salem, L.: Chem. Phys. Lett. *3*, 99 (1969).
[47] Bartell, L. S.: J. Chem. Educ. *45*, 754 (1969).
[48] Bader, R. F.: Mol. Phys. *3*, 137 (1960).
[49] — Can. J. Chem. *40*, 1164 (1962).
[50] Sevilla, M. O., Flajser, S. H., Vincow, G., Dauben, H. J.: J. Am. Chem. Soc. *91*, 4139 (1969).
[51] Longuet-Higgins, H. C., Pople, J. A.: Proc. Phys. Soc. (London) *A68*, 591 (1955).
[52] Toyota, A., Nakajima, T.: Chem. Phys. Lett. *6*, 144 (1970).
[53] Dewar, M. J. S., Gleicher, G. J.: J. Am. Chem. Soc. *87*, 3255 (1965).
[54] Allinger, N. L., Thai, J. C.: Theoret. Chim. Acta *12*, 29 (1968).
[55] Buenker, R. J., Peyerimhoff, S. D.: J. Chem. Phys. *48*, 354 (1968).
[56] Dewar, M. J. S., Kohn, M. C., Trinajstic, N.: J. Am. Chem. Soc. *93*, 3437 (1971).
[57] Gouterman, M., Wagnière, G.: J. Chem. Phys. *36*, 1188 (1962).

[58] Bregman, J., Hirshfeld, F. L., Rabinovish, D., Schmidt, G. M. J.: Acta Cryst. *19*, 227 (1965).
[59] Hirshfeld, F. L., Rabinovish, D.: Acta Cryst. *19*, 235 (1965).
[60] Blattmann, H. R., Heilbronner, E., Wagnière, G.: J. Am. Chem. Soc. *90*, 4786 (1968).
[61] Pariser, R., Parr, R. G.: J. Chem. Phys. *21*, 446, 767 (1953).
[62] Pople, J. A.: Trans. Faraday Soc. *49*, 1375 (1953).
[63] Dewar, M. J. S., Gleicher, G. J.: Tetrahedron *21*, 140 (1957).
[64] Mataga, N., Nishimoto, K.: Z. Physik. Chem. *13*, 140 (1957).
[65] Yamaguchi, H., Nakajima, T., Kunii, T. L.: Theoret. Chim. Acta *12*, 349 (1968).
[66] Fujimura, Y., Yamaguchi, H., Nakajima, T.: Bull. Chem. Soc. Japan *45*, 384 (1972).
[67] Hafner, K., Häfner, H., König, C., Kreuder, M., Ploss, G., Schultz, G., Sturm, E., Vöpel, K. H.: Angew. Chem. *75*, 35 (1963); Angew. Chem. Intern. Ed. Engl. *2*, 123 (1963).
[68] Nakajima, T., Saijo, T., Yamaguchi, H.: Tetrahedron *20*, 2119 (1964).
[69] Blood, C. T., Linstead, R. P.: J. Chem. Soc. *1952*, 2255, 2263.
[70] Chuen, C. C., Fenton, S. W.: J. Org. Chem. *22*, 1538 (1958).
[71] Yamaguchi, H., Terasaka, T., Nakajima, T.: Theoret. Chim. Acta *18*, 225 (1970).
[72] Lindner, H. J.: J. Chem. Soc. *1970*, B 907.
[73] Carstensen-Oeser, E., Habermehl, G.: Angew. Chem. *80*, 564 (1968).
[74] Quasba, R., Brandl, F., Hoppe, W., Huber, R.: Acta Cryst. *B25*, 1198 (1969).
[75] Yamaguchi, H., Nakajima, T.: Bull. Chem. Soc. Japan *44*, 682 (1971).
[76] Lo, D. H., Whitehead, M. A.: Chem. Commun. 771 (1968).
[77] Trost, B. M., Bright, C. M.: J. Am. Chem. Soc. *89*, 4244 (1967).
[78] Boekelheide, V., Vick, G. K.: J. Am. Chem. Soc. *78*, 653 (1956).
[79] Tajiri, A., Ohmichi, N., Nakajima, T.: Bull. Chem. Soc. Japan *44*, 2347 (1971).
[80] Pople, J. A., Santry, D. P., Segal, G. A.: J. Chem. Phys. *43*, S 129 (1965).
[81] — Segal, G. A.: J. Chem. Phys. *43*, S 136 (1965); *44*, 3289 (1966).
[82] Katz, T. J., Rosenberger, M.: J. Am. Chem. Soc. *84*, 865 (1962).
[83] — — O'Hara, R. K.: J. Am. Chem. Soc. *86*, 249 (1964).
[84] Hafner, K.: Angew. Chem. *75*, 1041 (1963); Angew. Chem. Intern. Ed. Engl. *3*, 165 (1964).
[85] Katz, T. J., Schulman, J.: J. Am. Chem. Soc. *86*, 3169 (1964).
[86] — Balogh, V., Schulman, J.: J. Am. Chem. Soc. *90*, 734 (1968).
[87] McConnel, H. M.: J. Chem. Phys. *24*, 764 (1956).
[88] Hunt, G. R., Ross, I. G.: J. Mol. Spectry. *9*, 50 (1962).
[89] Hoarau, J.: Ann. Chim. *1*, 544 (1956).
[90] Yamaguchi, H., Nakajima, T.: Unpublished work.
[91] Dauben, H. J., Wilson, J. D., Laity, J. L.: J. Am. Chem. Soc. *91*, 1991 (1969).
[92] Jung, D. E.: Tetrahedron *25*, 129 (1969).
[93] Mayot, M., Berthier, G., Pullman, B.: J. Phys. Radium *12*, 652 (1951); J. Chim. Phys. *50*, 176 (1953).
[94] Pullman, B., Pullman, A., Bergmann, E. D., Berthier, G., Fischer, E., Hirshberg, E., Pontis, J.: J. Chim. Phys. *49*, 24 (1952).
[95] Wagnière, G., Gouterman, M.: Mol. Phys. *5*, 621 (1962).
[96] Nakajima, T., Kohda, S.: Bull. Chem. Soc. Japan *39*, 804 (1966).
[97] Pople, J. A., Untch, K. G.: J. Am. Chem. Soc. *88*, 4811 (1966).
[98] Longuet-Higgins, H. C., in: Aromaticity. Special Publication No. 21, p. 109. London: The Chemical Society 1966.
[99] Van Vleck, J. H.: The theory of electronic and magnetic susceptibilities, Chap. VI. Oxford: The Clarendon Press 1932.
[100] Fernández-Alonso, J. I., Palou, J., in: Structural chemistry and molecular biology, p. 806. San Francisco: Freeman 1968.

Received January 17, 1972

Some Formal Properties of the Kinetics of Pentacoordinate Stereoisomerizations

Dr. Jean Brocas

Faculté des Sciences, Université Libre de Bruxelles, Brussels, Belgium

Contents

I. Introduction

In recent times, the study of molecular non-rigidity has been characterized by new and important developments in both experimental and theoretical directions. From the theoretical point of view, the pioneer work of Longuet-Higgins[1] initiated numerous further studies[2, 3]. The aim of this work was to extend the concept of symmetry to non-rigid molecules undergoing certain internal movements, the so-called feasible transformations. Some recent discussion of these problems may be found in Refs.[4, 5] for example.

Any molecule characterized by a number of configurations or by a wide variety of movements leading from one configuration to another challenges us to classify both configurations and internal movements. It is well known that cosets are associated with isomers or configurations[3] when the ligands on the molecular skeleton are all different. For other ligand partitions or for the classification of internal movements, double cosets[6] are an essential tool.

Non-rigidity has important consequences for the rotation and vibration spectra. Extensive experimental investigations exist in this domain[7]; they are based on the elaboration of model hamiltonians to describe the external motions. Recently, non-rigid molecule effects on the rovibronic levels of PF_5 have been examined[8], so leading to the prediction of the spectroscopic consequences of Berry[9] processes.

The pentacoordinate molecules of trigonal bipyramidal form, like PF_5, are a very nice example for the study of the formal properties of stereoisomerizations. They are characterized by an appreciable non-rigidity and they permit the description of kinetics among a reasonable number of isomers, at least in particular cases (see below). Therefore the physical and chemical properties of these molecules have been thoroughly investigated in relation to stereoisomerization. Recent reviews may be found in the literature on some aspects of this problem. Mislow[10] has described the role of Berry pseudorotation on nucleophilic addition-elimination reactions and Muetterties[11] has reviewed the stereochemical consequences of non-rigidity, especially for five- and six-atom families as far as their nmr spectra are concerned.

Since these coordination numbers give rise to numerous isomers and stereoisomerization processes, some effort has been necessary in order to represent them in a compact way. Graphs and matrices have been used extensively in this context.

In pentacoordinate chemistry, the Berry mechanism may be depicted by a graph which was first used in chemistry by Balaban[12]. Equivalent illustrations[13]–[16] of the topology of Berry steps have been provided in recent years and this technique has been extended to the description of other mechanisms[17] (for a recent review see[18]).

Even for pentacoordinate molecules, and a fortiori for hexacoordinate ones, graphs may become inadequate because they involve too many points and lines. Matrices have been used to overcome this difficulty[17], [19].

We think there is a need to investigate the algebraic properties underlying the various graph and matrix representations in order to discover the properties which are common to every stereoisomerization, to every ligand partition and perhaps, to every coordination number. This attempt is necessary if we want to understand the maximum number of phenomena with a minimum number of concepts.

More concretely, the aim of our investigation is to examine, from a theoretical point of view, the relation between the non-rigidity of pentacoordinate molecules and the characteristics of the temporal evolution of systems of such molecules towards chemical equilibrium. We also want to indicate the type of experimental information needed concerning the time evolution of these systems, in order to sharpen our ideas on the "feasibility"[1] of the internal movements. We here give an account of the main aspects of our attempt and try to present it in a unified and synthesizing fashion.

Three types of results have been obtained and may be summarized as follows.

First, we recall the five different stereoisomerization processes; these are necessary to reach any isomer from a given one in one step. However, they are not really independent because a succession of two processes is a linear combination of processes. A multiplication table of the processes has been established. This is explained in Sections II and III.

Second, the symmetry properties of one of the processes (the Berry step) are analysed. The operators associated with it are shown to commute with the elements of a cyclic group of order ten. Because of the structure of the multiplication table, the same is true for the operators associated with the other stereoisomerization processes. The solution of the rate equations for any process are derived from these properties (Sections IV and V).

These results are valid for the more *general* ligand partition: five different ligands. However, a chemical system of such a complexity is ill adapted for experimental investigation, so we must try to get information on molecules with less than five different ligands. The various possible *particular* ligand partitions (some ligands are the same) have been studied systematically and the formal results are summarized in Section VI. It appears that, for any particular ligand partition, the relation between the five processes, the rate equations and their solution may be obtained by means of very simple algebraic operations from the corresponding expressions for the general ligand partition.

We conclude Section VII with a suggestion of some experimental tests to yield information about the physical properties of the penta-coordinate molecules. The simplest test requires only equilibrium measurements: Our model assumes that the distribution of the ligands on the skeleton—for a given ligand partition—has no influence on the internal energy of the stable molecule nor of the intermediate states. Given this condition, it is easy to obtain theoretically the ratio of equi-librium isomer concentrations for each ligand partition. By compari-son with the experimental answer, one should be able to test the adequacy of the assumption for the chemical species under consideration. If the test is positive, it is worthwhile trying non-equilibrium measurements and determination of relaxation times. We indicate in Section VII the minimum ligand partition necessary to discriminate between the various stereoisomerization processes. If such an experimental investigation were possible, it should provide results which would enlarge our know-ledge of pentacoordinate stereoisomerizations.

We believe that the same type of description applies, with the same restrictions, to higher coordination numbers.

II. Description of Pentacoordinate Molecules

Let us examine the properties of molecules where a central atom M is surrounded by five ligands A, B, C, D, E. We assume that the ligands are at the vertices of a trigonal bipyramid. This assumption is adequate for most pentacoordinate complexes but we ougth to mention that the description of stereoisomerization we propose could be applied if another polytopal form—the tetragonal pyramid, for example—was the stable one. The same type of description has already been undertaken for hexacoordinate octahedral complexes[20].

It has been pointed out[6] that one must maintain a sharp distinction between the symmetry group of the skeleton (trigonal bipyramid) and the symmetry group of the ligands which can be defined as the permuta-tions of identical ligands among themselves and does not contain any reference to the skeleton. It depends exclusively on the *ligand partition.*

For the present case, we have seven such partions of the five ligands which may be symbolized by Young diagrams. The number of rows in such a graph is the number of sorts of ligands, and the length of each row is the number of ligands of a given sort. In Fig. 1, we recall the Young diagrams for the partition of five ligands. They have been "ordered" according to the criterion[21]: a diagram is smaller than another one if we can construct it from the other by pulling boxes from upper lines to lower ones.

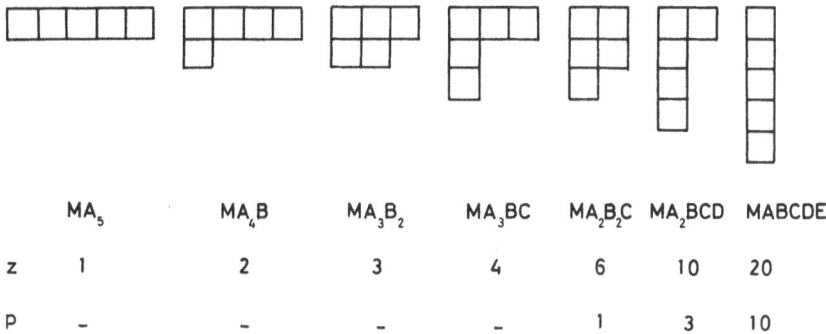

	MA_5	MA_4B	MA_3B_2	MA_3BC	MA_2B_2C	MA_2BCD	MABCDE
z	1	2	3	4	6	10	20
p	–	–	–	–	1	3	10

Fig. 1. Young diagrams for partitions of five ligands

For each ligand partition, it is interesting to give the number of possible isomers z and the number of enantiomeric pairs p. These numbers may be obtained merely by inspection[19] in this simple case. In general, they may be obtained from the relations between the symmetry group of the skeleton and the symmetry group of the ligand partition[6]. It is seen that three ligand partions are *active* (give rise to at least one enantiomeric pair), while the four others are *inactive* (no enantiomeric pair). It must be recalled that the ordering criterion is such that all partitions smaller than an active one, are active[21].

We will be interested in understanding some properties of the *rate equations* for stereoisomerization. We first recall previous descriptions of these stereoisomerizations[9,22] although we adopt a numbering which is more natural[23].

P_1: the Berry step is seen as a double bending of an equatorial and an apical angle. The two apical ligands become equatorial and two equatorial ones go to apical positions. One of the equatorial ligands, the so-called pivot, is on the fourfold axis of the tetragonal pyramidal intermediate state. The connectivity δ_1, i.e. the number of isomers reached from a given one in one step, is three.

J. Brocas

P_2: a twist permuting one apical and two equatorial ligands ($\delta_2=6$)
P_3: a twist permuting one apical and one equatorial ligand ($\delta_3=6$)
P_4: two simultaneous twists each involving one apical and one equatorial ligand ($\delta_4=3$)
P_5: a twist permuting two equatorial ligands ($\delta_5=1$)

These five processes have been defined independently of the ligand partition. They are visualized as properties of the skeleton symmetry only. In the coset[3] and double coset[6] formulations of stereoisomerization this idea is expressed in a precise mathematical form. The underlying assumption is that the presence of *different* ligands does not distort the skeleton geometry. It is certainly possible to find chemical situations where this is reasonably correct.

Let us now examine what happens for the more general ligand partition where there are five different ligands and twenty isomers.

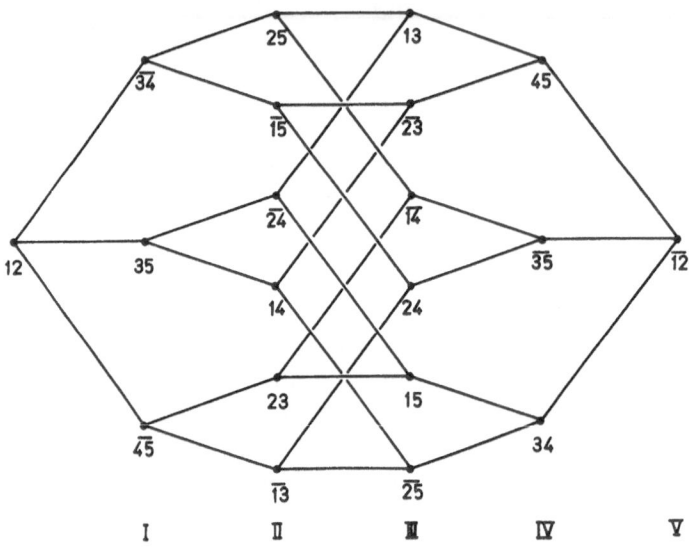

Fig. 2. Graph for pentacoordinate stereoisomerizations

Matrices and graphs have been used to describe the various processes[12)−19), 22]. As an example, we draw a graph for Berry steps in Fig. 2. The two-digit symbols are used according to the usual isomer numbering convention[22].

In this graph each point represents one isomer and each line a Berry step. In principle, each step happens with a different probability because the energies of the starting and final isomers and of the intermediate state are different: they depend on the positions of the ligands on the skeleton. However, as a first approximation, we will assume that the various probabilities for the various steps of a given process P_j all have the same value k_j. We thus neglect the influence of the ligand positions on the internal energy of the molecule. This amounts to a topological description of kinetics which is very close to Hückel's topological molecular orbital theory.

III. Formal Relations between the Five Processes

The five processes are not really independent. This is well illustrated in Fig. 2. The symbols I, II, III, IV, V under each group of isomers give the number of Berry steps needed to reach them from 12, considered as the starting isomer. However, if one performs one P_2 step, it is easy to convince oneself that one reaches also the six isomers of group II[24]. For this reason the symbols I, III, ... etc., refer also to the index of the process needed to reach the corresponding groups in one step. The only difference is that multiple steps allow for a wider variety of paths, because one can come back to the starting point.

For a process P_i, we may construct a matrix A_i where each row and column correspond to an isomer[17],[19]. If two isomers interconvert in one P_i step, we put a 1 at the intersection of the corresponding row and column. Such a matrix is associated[25] with the graph G_i for P_i. By performing directly the relevant matrix products or by counting the number of paths obtained by making one step on G_i followed by one step on G_j, it has been possible to establish[24] the product properties of the various operators corresponding to the stereoisomerizations. The structure of these relations is seen in the following expression:

$$A_i A_j = \sum_k c_{ij}^k A_k, \qquad (1)$$

which tells us that the succession of two processes i and j may be expressed as a linear combination of single processes k. The coefficient c_{ij}^k accounts for the number of times P_k appears in the succession $P_i P_j$. In (1) the index k varies from 0 to 5 ($A_0 \equiv E$). The fact that a relation like (1) exists is due to the fact that the six processes form a set which is *complete*[26]. From a given isomer it is always possible to reach anyone of them by using one of the steps because,

> a) two different processes acting on the same isomer give two different results, and
>
> b) the sum of the connectivities is equal to the number of isomers:

$$\sum_{i=0}^{5} \delta_i = 20. \tag{2}$$

The multiplication table for the processes has been given previously[24] and is repeated here for further reference (Table 1). It was

Table 1. Multiplication table for the processes

	E	A_1	A_2	A_3	A_4	A_5
E	E	A_1	A_2	A_3	A_4	A_5
A_1		A_2+3E	$2A_1+2A_3$	$2A_2+2A_4$	A_3+3A_5	A_4
A_2			$3A_2+4A_4+6E$	$3A_3+4A_1+6A_5$	$2A_2+2A_4$	A_3
A_3				$3A_2+4A_4+6E$	$2A_1+2A_3$	A_2
A_4					A_2+3E	A_1
A_5						E

obtained independently by Ruch and Hässelbarth, using double coset formulation[27]. Half of the table has not been written down because the operators corresponding to various processes commute

$$A_i A_j = A_j A_i \tag{3}$$

due to the fact that they are symmetrical and that their product (1) must be symmetrical too[24]. It must also be stressed that the structure of the multiplication table is more complicated than the structure of a group multiplication table because the product of two elements gives, not an element of the set, as it would in a group, but a linear combination of the elements of the set, which is in this sense complete.

It is clear from the table that every element may be expressed[20), 24)] in terms of A_1, A_5 and E, since one has

$$A_2 = A_1^2 - 3E, \tag{4}$$

$$A_4 = A_1 A_5, \tag{5}$$

$$A_3 = A_2 A_5 = (A_1^2 - 3E) A_5, \tag{6}$$

but it is more subtle to realize that A_5 itself "depends only" on A_1 and E. The argument is the following[28)]. One has

$$A_4 = \tfrac{1}{4}(A_2^2 - 3A_2 - 6E) \tag{7}$$

and, in the same way

$$A_1 A_2 = 2A_1 + 2A_3$$

yields

$$A_3 = \tfrac{1}{2}(A_2 A_1 - 2A_1). \tag{8}$$

But, since $A_4 A_1 = A_3 + 3A_5$, one may write

$$A_5 = \tfrac{1}{3}(A_4 A_1 - A_3) \tag{9}$$

and replace A_4 and A_3 by their values (7) and (8), which gives

$$A_5 = \frac{1}{3}\left[\frac{A_1}{4}(A_2^2 - 3A_2 - 6E) - \frac{1}{2}(A_2 A_1 - 2A_1)\right]$$

where (4) may be inserted. The result

$$A_5 = \frac{A_1}{12}\left[A_1^4 - 11A_1^2 + 22E\right] \tag{10}$$

shows that, because of (5) and (6), the operators A_2, A_3, A_4, and A_5, are functions of A_1 and E only.

Suppose now that we could find a similarity transformation such that $N^{-1} A_1 N = \tilde{A}_1$ should be a diagonal matrix. It is then obvious that A_2, A_3, A_4 and A_5 being simple polynomials in A_1, should be diagonal, too, because

$$N^{-1} A_1^n N = N^{-1} A_1 N N^{-1} A_1 N \dots N^{-1} A_1 N = \tilde{A}_1^n \tag{11}$$

which is diagonal. We will indicate in the next section how this similarity transformation can be found and how it leads to the solution of kinetic equations for every stereoisomerizations.

IV. Symmetry Properties of Berry Processes

The symmetry properties of the pentacoordinate stereoisomerizations have been investigated[29] on the Berry processes. They have been analyzed by defining two operators Q and I[30]. The operator I is the geometrical inversion about the center of the trigonal bipyramid. Since this skeleton has no inversion symmetry, I moves the skeleton into another position. Moreover, if the five ligands are different, it transforms any isomer into its enantiomer, as shown in Fig. 3.

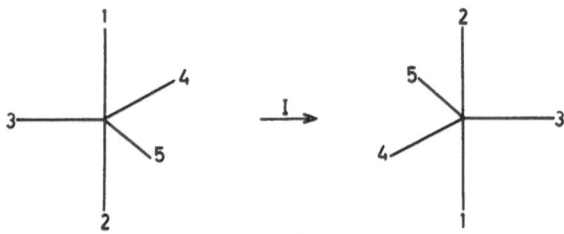

Fig. 3. The action of I on (12)

The other operator is a cyclic permutation $Q = (13\,542)$ on the five ligands, where $(d_1 d_2 d_3, \ldots, d_n)$ means replace object d_1 by d_2, d_2 by $d_3, \ldots, d_n$ by d_1. The use of the standard rules for calculating products of permutations yields:

$$Q^2 = (15\,234); \quad Q^3 = (14\,325); \quad Q^4 = (12\,453); \quad Q^5 = E \quad (12)$$

where E is the identity operation. It is easy to verify that Q and I commute

$$QI = IQ \equiv R \tag{13}$$

and R may be taken as the generator of the cyclic group of order ten whose elements are R^p $(1 \leqslant p \leqslant 10)$. Table 2 shows the effect of R^p on isomer 12. In this way the ten isomers of the first line may be generated. The second line is generated in a parallel way by starting from 34 or from any isomer absent in the first line. The action of R does *not* enable us to go from one line to the other. We will classify the twenty isomers in four families (I, II, $\bar{I}$, $\bar{II}$) as indicated in Table 2[29].

Table 2. Action of $R = Q \cdot I$

	E	Q	Q^2	Q^3	Q^4	I	IQ	IQ^2	IQ^3	IQ^4
I	12	$\overline{13}$	$\overline{35}$	45	$\overline{24}$	$\overline{12}$	13	35	$\overline{45}$	24	$\overline{I}$
II	34	25	14	23	$\overline{15}$	$\overline{34}$	$\overline{25}$	$\overline{14}$	$\overline{23}$	15	$\overline{II}$

In the vectorial space defined previously[29] one may represent each operator by a twenty-by-twenty matrix. This, is true for E, Q, I, R but also for the operators A_1, A_2, A_3, A_4 and $A_5 = I$. Except for the order of rows and columns, they are identical to the matrices associated to the graphs G_i. In these notations, the rate equations for the five processes are written:

$$\frac{dC}{dt} = k_i(A_i - \delta_i E) \cdot C. \qquad (14)$$

C, the concentration vector, being a column vector of dimension twenty and i varying from 1 to 5. The neglect of the influence of the ligand positions on the internal energy of the molecule is already contained in the above expression.

Using the explicit expressions for R and A, the following relation has been established[29] for Berry processes:

$$R^{-1} A_1 R = A_1 \qquad (15)$$

which means that A_1 is invariant with respect to the transformation R or that A and R commute. More precisely, the similarity transformation R in the l.h.s. of (14) reorganizes merely the rows and columns of A_1, since R contains one and only one non-vanishing element in each row or column. The effect of this reorganization is to provide a matrix which is identical to the starting one. The property (15) remains true for every power R^p and, as a result, A_1 is invariant with respect to any transformation R^p of the cyclic group of order ten. For well-known group theoretical reasons, A_1 must be block-diagonal in the irreducible representations of the cyclic group of order ten or, in other words, in the space of the eigenvectors of R.

These ten eigenvectors are easy to write down. Each of them must be used twice in the twenty-dimensional space, once for families I and $\bar{I}$, once for families II and $\overline{II}$, and in this manner the matrix M can be constructed[29] such that

$$M^{-1} A_1 M = A_1'$$ (16)

where A_1' has ten two by two blocks on its main diagonal, because each of the one-dimensional representations of the cyclic group of order ten appears twice in A_1. From the knowledge of the blocks and of the eigenvectors of R, one deduces the eigenvectors and the eigenvalues of A_1. The eigenvectors of A_1 are linear combinations of one eigenvector of R belonging to subspace I plus $\bar{I}$ and one similar eigenvector belonging to subspace II plus $\overline{II}$. The explicit form of the eigenvectors of A_1 may be found elsewhere[29]. The matrix N diagonalizing A_1 is obtained in the usual manner:

$$N^{-1} A_1 N = \tilde{A}_1 \quad \text{(diagonal matrix)}.$$ (17)

V. Solution of the Rate Equations

Of course, the solution of the rate equations (14) follows immediately from the knowledge of N and from the fact that N diagonalizes every matrix $A_i\ (i=1,...,5)$. Using the notations defined previously[24], one may recall the results:

$$\gamma_r(t) = \gamma_r(0)\, e^{b_i^{(r)} t}$$ (18)

with

$$b_i^{(r)} = k_i\, (a_i^{(r)} - \delta_i)$$ (19)

is the rth eigenvalue of the operator $B_i = k_i\, [A_i - \delta_i E]$ describing the stereoisomerization P_i. Moreover γ is the column vector whose twenty rows $(r=1,...,20)$ contain the chemical normal modes and defined by

$$\gamma = N^{-1} C.$$ (20)

From these expressions, the explicit time dependence of each concentration has been obtained[24]. The concentration of a given isomer at time t is a linear combination of the concentrations at time zero, the coefficients are themselves linear combinations of the time dependent exponentials $e^{b_i^{(r)} t}$.

The corresponding formulae have been listed in previous work[24]. They will not be reproduced here because we think that the most characteristics of the result can be discussed on the spectrum of inverse relaxation times given in Table 3.

From this table, it is seen that the eigenvalues of $\boldsymbol{B}_i$ or, equivalently, the inverse relaxation times are in our description integer multiples of the unknown k_i. The point we want to stress is that these integers are *different* for each process P_i so that, even if the five rate constants k_i

Table 3. The five spectra of relaxation times

Values of r	1		2		3		4		5	
	$a_1^{(r)}$	$b_1^{(r)}$	$a_2^{(r)}$	$b_2^{(r)}$	$a_3^{(r)}$	$b_3^{(r)}$	$a_4^{(r)}$	$b_4^{(r)}$	$a_5^{(r)}$	$b_5^{(r)}$
1	3	0	6	0	6	0	3	0	1	0
8, 12, 16, 20	2	$-k_1$	1	$-5k_2$	-1	$-7k_3$	-2	$-5k_4$	-1	$-2k_5$
2, 5, 9, 13, 17	1	$-2k_1$	-2	$-8k_2$	-2	$-8k_3$	1	$-2k_4$	1	0
4, 7, 11, 15, 19	-1	$-4k_1$	-2	$-8k_2$	2	$-4k_3$	1	$-2k_4$	-1	$-2k_5$
6, 10, 14, 18	-2	$-5k_1$	1	$-5k_2$	1	$-5k_3$	-2	$-5k_4$	1	0
3	-3	$-6k_1$	6	0	-6	$-12k_3$	3	0	-1	$-2k_5$

were the *same*, the time evolution of the concentrations would be different for each process. These important differences in the spectra of relaxation times reflect the topological differences between the kinetic scheme of each process. They could be investigated in relaxation experiments. We come back to that point in the next section.

From the point of view of the kinetic distinction between the various processes, we think that one should proceed in two steps:

1) It is relatively easy to distinguish from each other processes of different connectivity: P_5 from P_1 and P_4, from P_2 and P_3.

2) It is more difficult, but not impossible (see below), to distinguish between processes having the same connectivity: P_1 from P_4; P_2 from P_3. This distinction rests on the existence of chemical normal modes corresponding to the eigenvalue -1 for P_5 (second, fourth and sixth line of Table 3). Only these ten modes distinguish P_1 from P_4 and P_2 from P_3. This may be easily understood by comparison with formulae (5) and (6).

VI. Particular Ligand Partitions

In the three preceding sections, we have described some formal results valid for the more general ligand partition: five different ligands, last column of Fig. 1. We now give an account of some properties which are typical for the other ligand partitions, where we have less that five different ligands. We only give a qualitative description. Details and precise mathematical formulations are given elsewhere[20), 31)].

If one starts from the general ligand partition, each particular partition may be obtained by putting some of the ligands equal to each other. Such a *particularization* induces the indistinguishability of some of the isomers, which were all distinguishable in the general ligand partition. In this way the twenty isomers are partitioned into z classes, the kth class containing D_k isomers (which are distinguishable in the general ligand partition and indistinguishable in the considered partition). One has, of course

$$\sum_{k=1}^{z} D_k = 20. \tag{21}$$

We will use constraint matrices[17)] to describe the classification of the isomers into various classes: a given ligand partition is characterized by a matrix where each row is associated with a class and each column with an isomer in the general ligand partition. Such a matrix Θ has therefore z rows and twenty columns. In a given row, we put one in the column of the isomers belonging to the class of the row. Of course, there is one and only one non-vanishing element in each column, and D_k in the row k. By usual matrix calculation, it is easy to show that

$$\Theta \Theta_T = D \tag{22}$$

is a diagonal matrix with z rows and z columns whose elements are D_k (Θ_T means transposition of Θ).

The constraint matrices for typical ligand partitions may be found elsewhere[17)]. We have investigated some of their formal properties, in relation to stereoisomerization kinetics[20)]. We give an account of the main results. Details and demonstration have been worked out in reference[31)].

The stereoisomerization kinetics of the various processes and for particular ligand partitions may be described in two essentially different ways:

i) Since the five processes may be performed independently of the ligand partition, let us imagine that they occur on molecules where some of *the ligands have already been made equal*. We may then draw graphs where each point now represents a class. From *a given isomer* of class k, we may, using P_i, reach δ_i, isomers belonging to the classes l, l', l'' and so on (it could well be that one of the l's is k itself). Let $(\alpha_i)_{kl}$ be the number of isomers of class l reached by one p_i step from one isomer of class k. On the graph, we decide to draw, in this case, $(\alpha_i)_{lk}$, directed lines, with an arrow pointing from k to l. Proceeding in the same way, taking every time one isomer from each class, we obtains a multigraph[25] which is characteristic of process p_i and of the ligand partition. Examples of such graphs may be found in reference[23] but for our purpose we do not need, the explicit specification of the intermediate state which is made in this work. Other examples have been described[20] in connection with the solution of the rate equations for stereoisomerization with particular ligand partitions (see below). A matrix α_i may be associated with such a multigraph and, we have of course[20]

$$\sum_{l=1}^{z} (\alpha_i)_{lk} = \delta_i \tag{23}$$

expressing the fact that the total number of isomers starting from one isomer of any class k must be equal to the connectivity of P_i.

ii) We may also adopt the opposite point of view and consider the product ΘA_i, which means that we examine *first* the action of P_i on a given isomer of the general ligand partition (A_i) and that afterwards we put some ligands equal to each other (Θ). [In i) kinetics occurred *after* particularization.] But we expect that these two operations have to *commute*. This important property will not be demonstrated here, but may be suggested by the following argument. Suppose that a process P_i transforms an isomer a into δ_i other isomers $a', a'',\ldots$, etc. when the ligand partition is the general one. Let us now put some of the ligands equal to each other so that a, a', a'', become b, b', b'' (the a's are different from each other, but some of the b's may be equal).

The property tells us that the operation P_i, expressed by α in the particular ligand partition, performed on b yields $b' b' b'' \ldots$.

This commutation theorem may be stated in the following form

$$\Theta A_i = \alpha_i \Theta \tag{24}$$

and is symbolized graphically in Fig. 4. An important consequence of this relation may be obtained by multiplying it to the right by Θ_T and using (22). We get:

$$\alpha_i = \Theta A_i \Theta_T D^{-1} \tag{25}$$

since D is diagonal. This expression is quite general. From the knowledge of the matrix A_i for any process and for the general ligand partition, it provides the matrix α_i for the same process and any particular ligand partition. The validity of (25) has been tested in all possible cases for pentacoordinate stereoisomerization[31]. There is no doubt that it is also valid for other coordination numbers.

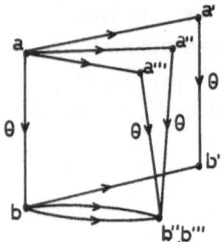

Fig. 4. The commutation property

It has also been shown[31] that the concentration vector for a particular ligand partition is given by $\boldsymbol{\Theta C}$. It is a column vector whose components are the z sums of the concentrations of the isomers belonging to the same class. It is easy to put (14) into the form

$$\frac{d(\boldsymbol{\Theta C})}{dt} = k_i(\boldsymbol{\alpha}_i - \delta_i \boldsymbol{E}_z)\boldsymbol{\Theta C} \tag{26}$$

which are the rate equations for process P_i in a particular ligand partition ($\boldsymbol{E}_z$ = unit matrix in z dimension).

The set of relations (1) have been shown[31] to hold for every particular ligand partition

$$\alpha_i \alpha_j = \sum_k c_{ij}^k \alpha_k \tag{27}$$

with the same coefficients c_{ij}^k. It indicates that the relations between the five processes remain the same, independently of the ligand partition. This is due to the fact that they are properties of the skeleton.

To close the list of formal properties of the kinetic equations for stereoisomerizations with particular ligand partitions, let us simply recall that the solution[24] of process P_i

$$C(t) = N e^{N^{-1}A_i N t} N^{-1} C(0) \tag{28}$$

and for the general ligand partition may be put into a form adapted for particular ligand partition[31]

$$\boldsymbol{\Theta} C(t) = v e^{v^{-1}\alpha_i v t} v^{-1} \boldsymbol{\Theta} C(0) \tag{29}$$

where v is the matrix of the eigenvectors of α_i. Each eigenvector of α_i may in fact be obtained through the action of Θ on an eigenvector of A_i. It must also be stressed that both of them correspond to the same eigenvalue. As a consequence, the eigenvalues of α_i belong to the set of eigenvalues of A_i. However, some eigenvalues of A_i do not appear in α_i or appear in it with reduced degeneracies. The rules underlying these facts have been stated elsewhere[31].

VII. The Possibility of Experimental Investigation

The possibility of experimental investigation is, of course, restricted to systems with particular ligand partition, where the number of isomers is not too big. The aim of such investigations is to obtain information about the relative probabilities of the five mechanisms which have been defined, till now, on symmetry grounds only. We think that a comparison of the experimental spectra of relaxation times with the various possible theoretical ones (see Table 3) should furnish answers to that question, provided the assumptions of the model are a good description of the physical system.

There exists a very simple verification of this hypothesis. Indeed, if the internal energy of the molecule does not depend on the distribution of the ligands on the molecule, then—as we have assumed throughout this work—every step of a given process P_i has the same probability k_i. But, in this case, it is possible to show[31] that the ratio of equilibrium concentrations of the isomers in the particular ligand partition is equal to the ratio of the number of isomers in the corresponding classes of the general ligand partition:

$$\frac{c_k^{eq}}{c_l^{eq}} = \frac{D_k}{D_l} \tag{30}$$

and, if the above assumption is realistic, the experimental value of the equilibrium ratios should be reasonably close to the theoretical value (30). If this test is positive, it is worthwhile examining the properties of the spectrum of relaxation times.

On Table 3, it is clear that the spectra of relaxation times differ from one process to the other. But, if we progressively make the ligands equal to each other going from the partition MABCDE of Fig. 1 to the partition MA_5, we expect that the differences between the relaxations due to the various processes will vanish progressively (for MA_5

there is no relaxation at all, for any of the five processes). We have examined[31] the theoretical spectra of relaxation times for each ligand partition, in order to determine the minimum ligand partition where two given processes remain distinguishable. We do not list the spectra here, but merely formulate some conclusions.

i) The ligand partitions in the two first columns of Fig. 1 tell us nothing about relaxation. The first one, MA_5 because there is only one isomer and no relaxation at all. The second one, MA_4B because there is only one relaxation mode with a decay time which is five times the rate constant of the process under consideration. This negative conclusion about MA_4B is not irreconcilable with the fact that Whitesides and Mitchell[32] have distinguished for such a molecule between the mechanisms of connectivity three (P_1 and P_4) and those of connectivity six (P_2 and P_3) because an nmr experiment distinguishes axial and equatorial A's and because their molecule satisfies the supplementary condition that B remains equatorial.

ii) If the ligands have small electronegativity differences, such a condition is not expected to hold and, in that case, MA_3B_2 is the minimum ligand partition where the spectra of connectivity three is distinguishable from the spectra of connectivity six. For that partition, P_1 and P_4 are both characterized by relaxation times which are twice and five times their respective rate constants; for P_2 and P_3, the relaxation times are eight and five times the rate constants. Since this ligand partition gives rise to only three isomers, relaxation experiments on this system should not be too difficult.

iii) The distinction between processes of the same connectivity (P_1 from P_4; P_2 from P_3) is more subtle and requires at least MA_2B_2C as ligand partition. In this case there are six isomers. The reason why such a complicated system is needed is easy to understand: it corresponds to the minimum active partition of the ligands and, as seen on Table 3, the differences between the relaxation of P_1 and P_4 or between P_2 and P_3 exist only for active relaxation modes (those for which the eigenvalue of $P_5 = I$ is -1).

To conclude, we think that valuable information can ce obtained from such relaxation experiments. They could provide a direct, kinetic proof of the conjecture that the Berry mechanism is the most probable one, as is indicated by some recent experimental[18), 32)] and theoretical[33] work. The applicability of this model is however restricted to situations where the energy of the molecule does not depend on the distribution of the ligands on the skeleton and where, as a consequence, there is one rate constant for each process. If this is not true, the present description could be the first-order approximation of a perturbation calculation. Such a work will be undertaken soon.

Acknowledgements. This work has been performed in the Department of Professor Prigogine. We thank him for his encouragements and his interest.

Our collaboration with the "Collectif de Chimie Organique Physique" has been exceptionnaly useful for us. We want to acknowledge many discussions with Professors J. Nasielski and M. Gielen and Mr. R. Willem.

Several aspects of this work have also been discussed with Professors J. Philippot and J. M. Gilles, we thank them cordially.

References

[1] Longuet-Higgins, H. C.: Mol. Phys. *6*, 445 (1963).
[2] Altmann, S. L.: Proc. Roy, Soc. (London) *A 298*, 184 (1967).
[3] Watson, J. K. G.: Can. J. Phys. *43*, 1996 (1965).
[4] Altmann, S. L.: Mol. Phys. *21*, 577 (1971).
[5] Watson, J. K. G.: Mol. Phys. *21*, 587 (1971).
[6] Ruch, E., Hässelbarth, W., Richter, B.: Theoret. Chim. Acta. *19*, 288 (1970).
[7] See for example Bauder, A., Mathier, E., Meyer, R., Ribeaud, M., Günthard, Hs. H.: Mol. Phys. *15*, 597 (1968). – Mathier, E., Welti, D., Bauder, A., Günthard, Hs. H.: J. Mol. Spectry. *37*, 63 (1971).
[8] Dalton, B. J.: J. Chem. Phys. *54*, 4745 (1971).
[9] Berry, R. S.: J. Chem. Phys. *32*, 933 (1960).
[10] Mislow, K.: Acc. Chem. Res. *3*, 321 (1970).
[11] Muetterties, E. L.: Acc. Chem. Res. *3*, 266 (1970).
[12] Balaban, A. T., Farcasiu, D., Banica, R.: Rev. Roumaine Chim. *11*, 1205 (1966).
[13] Dunitz, J. D., Prelog, V.: Angew. Chem. (Intern. Ed.) *7*, 726 (1968).
[14] Lauterbur, P. C., Ramirez, F.: J. Am. Chem. Soc. *90*, 6722 (1968).
[15] Gielen, M., Nasielski, J.: Bull. Soc. Chim. Belges *78*, 339 (1969).
[16] Debruin, K. E., Nauman, K., Zon, G., Mislow, K.: J. Am. Chem. Soc. *91*, 7031 (1969).
[17] Gielen, M.: Med. Vlaamse. Chem. Ver. *31*, 185 (1969).
[18] Gielen, M.: Application of graph theory to the problem of pseudorotation in metall-organic compounds. In: Applications of graph theory to chemistry (ed. A. T. Balaban). New York: Academic Press (to appear 1972).
[19] Muetterties, E. L.: J. Am. Chem. Soc. *91*, 4115 (1969).
[20] Willem, R.: Mémoire de Licence. Univ. Libre de Bruxelles 1971.
[21] Ruch, E.: Acc. Chem. Res. (to be published).
[22] Muetterties, E. L.: J. Am. Chem. Soc. *91*, 1636 (1969).
[23] Gielen, M., Brocas, J., De Clercq, M., Mayence. G., Topart, J., in: Proceedings of the third Symposium on Coordination Chemistry (ed. M. T. Beck), Vol. 1, p. 495. Akadémiai Kiadó (Publishing house of the hungarian Academy of Science) 1970.
[24] Brocas, J., Willem, R.: Theoret. Chim. Acta I (to appear 1972).
[25] Berge, C.: La théorie des graphes et ses applications. Dunod 1963.
[26] Gielen, M., Van Lautem, N.: Bull. Soc. Chim. Belges *79*, 679 (1970).
[27] Ruch, E., Hässelbarth, W.: Personnal communication (1971).
[28] Gilles, J. M.: Personnal communication (1971).
[29] Brocas, J.: Theoret. Chim. Acta *21*, 79 (1971).
[30] Brocas, J., Gielen, M.: Bull. Soc. Chim. Belges *80*, 207 (1971).
[31] Brocas, J., Willem, R.: Theoret. Chim. Acta II (to appear 1972).
[32] Whitesides, G. M., Mitchell, H. L.: J. Am. Chem. Soc. *91*, 5384 (1969).
[33] Tee, O. S.: J. Amer. Chem. Soc. *91*, 7144 (1969).

Received February 10, 1972.

Radiochemical Transformations and Rearrangements in Organometallic Compounds

Prof. Dr. Donald R. Wiles

Chemistry Department, Carleton University, Ottawa, Canada

Prof. Dr. Franz Baumgärtner

Fachbereich Physikalische Chemie der Universität Heidelberg und Institut für Heisse Chemie der Gesellschaft für Kernforschung, Karlsruhe

Contents

I. Introduction

The first paper describing the effect of nuclear transformation on an organometallic compound was that of Mortenson and Leighton[1] who observed that a high proportion of $^{210}Bi(CH_3)_3$ was formed from the beta decay of $^{210}Pb(CH_3)_4$. Since that time a good deal of work has been published in this field; the subject has not, however, been properly covered in a review, and the field seems to lack coherence. This general area is of special interest, since it represents a middle ground between the Szilard-Chalmers reactions in ionic compounds and those in organic compounds. The radiochemically interesting atom lies at the centre of the molecule, and hence these compounds are geometrically similar to many of the ionic compounds. Chemically, however, they are more like the organic compounds.

We are thus in this review* concerned with compounds having metal-to-carbon covalent bonds—whether these be σ, π, or any other form of covalent bonding. Not included are discussions of cyanide complexes or of the many complex ions whose bonding is through nitrogen, oxygen or sulphur, in which the processes of recombination seem to be more like those of ionic solids than like those of the molecular organometallic compounds. The compounds of interest are mostly solids, with some liquids and a few gases, all of which are composed of discrete molecules having no strong electronic interaction with one another.

Generally speaking, radiochemical studies with organometallic compounds, as with most other compounds, have been undertaken with one or more of the following objectives:

> Isotope enrichment,
> molecular synthesis,
> mechanistic investigation.

Isotope enrichment is usually directed simply at producing radioisotopes of high specific activity. The work of Herr and his coworkers[2] is dominant among early studies in this field, with their investigations of many metal phthalocyanines. Among metal-carbon bonded compounds, the earliest work is that of Melander[3], who found enrichment factors of greater than 100 for antimony-124 in neutron-irradiated triphenylstibine. The difficulty with this method of preparing carrier-free radioisotopes on a commercial scale is that the compounds are usually sensitive to radiation damage. Thus, with the long irradiations

---

* We have attempted to cover the published literature up to mid-1971.

necessary to produce significant quantities of activity, the compounds tend to decompose seriously, with the resulting decrease in specific activity. These problems can usually be overcome by striking a balance between yield and specific activity, as is shown in a recent study of the irradiation of gallium phthalocyanine[4] and others[5]–[8], although the method is far from ideal.

In some cases the isotope enrichment is for the purpose of isolating and studying the nuclear properties of the nuclides, particularly when very selective or rapid separation methods may be needed. One of the earliest such works is that of Maurer and Ramm[9], who studied the decay modes of ^{209}Pb by working with gaseous $Pb(CH_3)_4$ and $Bi(CH_3)_3$, whose decomposition products following beta decay were collected and measured without interference from the many other activities usually present. Other interesting applications of a similar nature have been published[10]–[16]. Notable among these is Starke's use[11] of uranyl benzoylacetonate for enrichment of (n, γ) produced ^{239}U and the subsequent isolation of ^{239}Np. In a somewhat later work[12]–[15], a series of short-lived decay chains starting with ^{103}Mo, ^{104}Mo and ^{105}Mo were studied by catching fission recoils in $Cr(CO)_6$ so as to produce $Mo(CO)_6$ which could then quickly and easily be separated from the fission mixture by sublimation. Table 1 gives a list of isotope separation and enrichment studies which have been done by Szilard-Chalmers methods using metalorganic compounds.

Molecular synthesis by radiochemical means[17]–[20] has taken two forms, the most obvious of which is the recoil synthesis of labelled molecules for tracer purposes. It is indeed surprising that this method has not more often been used to synthesize tracer molecules for biological or medical purposes. Not only is the specific activity extremely high in many cases, but also molecules can be synthesized by these methods which can be prepared by no other means on a small scale. Here we refer, of course, to molecules which are not simply isotopic with the target compound. Perhaps the first such synthesis of biologically important metalorganic molecules is that of Hoi and coworkers[31], who investigated the indirect preparation of various radioarsenicals, using high specific activity ^{76}As prepared by the Szilard-Chalmers reaction on $AsCl_3$. Direct recoil synthesis of phenylmercury compounds has been done by Heitz and Adloff[32]. A similar synthesis has been reported by Wheeler and McClin[33].

The other form taken by molecule syntheses in the present context is that of synthesis of novel molecules which had not previously been known[19], or in any case are prepared only for the study of these molecules themselves. Here we refer to the synthesis of such molecules as $Tc(C_6H_6)_2^+$ [34],[35] $Rh(C_5H_5)_2$[36] and others, all by beta decay of a pre-

Table 1 a. Recoil separations for nuclear studies

Nuclide	Target	Reference
^{76}As	$(CH_3)_2AsCOOH$ (n,γ)	10
^{239}Np	Uranium benzoyl-acetylacete (n,γ)	11
^{209}Pb	$Pb(CH_3)_4$	9
^{104}Mo-^{104}Tc	$UO_3 + Cr(CO)_6$ (n,f)	12
^{105}(Mo, Tc, Ru)	$UO_3 + Cr(CO)_6$ (n,f)	13
^{103}(Mo, Tc, Ru)	$UO_3 + Cr(CO)_6$ (n,f)	14
$^{103, 104, 105}$Tc	$UO_2(CH_3COO)_2 + Cr(CO)_6$ (n,f)	15
^{134}Te	$UO_3 + Sn(C_6H_5)_4$ (n,f)	16

Table 1 b. Recoil separations for isotope production or enrichment

Nuclide	Target[a]	Enrichment factor	Reference
$^{122, 124}$Sb	$(C_6H_5)_3Sb$	100	3
$^{122, 124}$Sb	$(C_6H_5)_3Sb$	200	21
$^{122, 124}$Sb	$(C_6H_5)_3Sb$	500	22
Zn, Ga, In, V, Mo, Pd, Os, Ir, Pt	Metal Phthalocyanines	1000	2
Sn	$(C_6H_5)_4Sn$	2800	23
^{69}Ge	$(C_6H_5)_4Ge$ (γ,n)	10^3-10^4	24
^{122}Sb	$(C_6H_5)_3SbCl_2$ (γ,n)	10^3-10^4	
^{74}As	$(C_6H_5)_3AsCOOH$ (γ,n)	10^3	6
	$(C_6H_5)_3As$ (γ,n)	10^3-10^4	
^{51}Cr	$Cr(CO)_6$	10^4	18, 25
^{210}Bi	$(C_6H_5)_3Bi$	500	26
$^{75, 77}$Ge	$(C_2H_5)_4Ge$		27
^{99}Mo	$Mo(CO)_6 +$ Oxalic acid	500	8, 28
$^{197, 197m, 203}$Hg	R_2Hg ($R = CH_3, C_2H_5,$ or C_6H_5)	10^2-10^3	7, 29
^{51}Cr	$Cr(CO)_6$	10^4	5
^{99}Mo	$Mo(CO)_6$	260	
^{187}W	$W(CO)_6$	130	

[a] Bombardment is with neutrons to give (n,γ) unless otherwise noted.

cursor molecule whose synthesis is straightforward. This too, is a process which could be more widely employed to produce at least tracer quantities of inaccessible compounds, inasmuch as the ligand structure of the molecule seems not to be disturbed by beta decay. Table 2 gives a list of metalorganic compounds which have been synthesized in this and in similar ways.

Mechanistic investigations. It is perhaps an exaggeration to describe the studies to date as "mechanistic". The phenomena which occur are so complex, and vary so much from one situation to another that most of the

Table 2. Synthesis by recoil methods: a partial list of radioactive organometallic compounds, other than the starting material, synthesized by recoil methods

Molecule	Method	Reference
$^{99m}Tc(C_6H_6)_2^+$	$^{99}Mo(C_6H_6)_2 \xrightarrow{\beta}$	34, 37
$^{103}RuCp_2$	$U_3O_8 + FeCp_2\ (n,f)$	37, 38, 39
$^{99}Mo(CO)_6$	$U_3O_8 + Cr(CO)_6\ (n,f)$	40
$Cp^{99}Tc(CO)_3$	$[Cp^{99}Mo(CO)_3]_2 \xrightarrow{\beta}$	41
$^{105}RhCp_2$	$^{105}RuCp_2 \xrightarrow{\beta}$	36
$^{151}PmCp_3$	$^{151}NdCp_3 \xrightarrow{\beta}$	42
$Te\phi_2$ $Sb\phi_3$	$Sn(C_6H_5)_4 + UO_2\ (n,f)$	16
$^{56}Mn(CO)_5$	$Mn_2(CO)_{10}\ (n,\gamma)$	43
$CH_3{}^{56}Mn(CO)_5$	$CH_3CpMn(CO)_3\ (n,\gamma)$	44
$H^{56}Mn(CO)_5$	$CH_3CpMn(CO)_3\ (n,\gamma)$	45
$^{56}Mn_2(CO)_{10}$	$CpMn(CO)_3\ (n,\gamma)$	45
$Cp^{56}Mn(CO)_3$	Fulvalene $(Mn(CO)_3)_2\ (n,\gamma)$	46
$(C_6H_6)_2{}^{51}Cr$ $^{51}Cr(CO)_6$	$C_6H_6Cr(CO)_3\ (n,\gamma)$	47
$\phi^{56}Mn(CO)_5$	$Mn_2(CO)_{10}$ in C_6H_6	45
$As\phi_3$	$AsCl_3$ + benzene	48
$[(CH_3)_2C_6H_3]_2{}^{210}Po$	$^{210}Bi[(CH_3)_2C_6H_3]_3 \xrightarrow{\beta}$	49
$^{210}Po(C_6H_5)_2Cl_2$	$^{210}Bi(C_6H_5)_3Cl_2 \xrightarrow{\beta}$	50, 51
$^{210}Bi(C_6H_5)_3Cl_2$	$^{210}Pb(C_6H_5)_3Cl \xrightarrow{\beta}$	52
$^{210}PoAr_3X$	$^{210}BiAr_3 \xrightarrow{\beta}$ $Ar = \phi CH_3, \phi(CH_3)_2$	49
$^{103}RuCp_2$	U in Cp (n,f)	53
ϕ_2Hg $\phi HgCl$	$\phi HgCl\ (n,\gamma)$ $\phi_2Hg\ (n,\gamma)$	33

work to date has been quite rightly directed toward simply finding general patterns of behaviour. In this sense, the mechanistic study of radiochemical transformations in metalorganic compounds began in 1953, with the papers by Edwards, Day and Overman[54] and Maddock and Sutin[55] on tetramethyl lead and triphenylarsine, respectively. Studies with this main objective are the subject of the present review.

II. Survey of Experimental Results

A. Processes with No Change in Atomic Number

In this group the most commonly used reaction is that of radiative neutron capture. Also here are to be found $(n, 2n)$ and (γ, n) reactions, although very few studies have been done with these reactions, and isomeric transitions (although these may often be more profitably discussed along with electron capture reactions).

1. Metal Alkyls and Aryls

Studies of metal alkyls and aryls deal largely with arsenic, germanium and antimony, although some work has also been done on tellurium, mercury, thallium, bismuth and lead. The major contributions can be neatly divided into four periods: early studies by Maddock, Sutin and Hall[55), 56), 57)], studies by the Polish[48), 58), 59)] and the Strasbourg[29), 32), 79)] groups, work by Riedel and Merz[30), 60)–64)], and most recently a series of investigations by Grossmann[66)–75)] and by Nowak and Akerman[76)]. The early work suffered (as has so much subsequent work) from what Maddock and Sutin themselves recognized to be a poor separation method. Column chromatography with a series of successively more polar solvents was able to effect separation of up to ten distinct fractions—not all of which were pure, and not all of which could be adequately identified. Despite this handicap, Maddock and Sutin were able to establish the final identity of about 85% of the ^{76}As produced in neutron-irradiated triphenylarsine. Since the results were dependent on the exact experimental conditions, suffice it to say that a large fraction of the radioarsenic atoms were bonded to at least one phenyl group. Thermal annealing effects were also evident, and even as low as 45 °C the yield of Asϕ_3 doubled in 30 h, evidently at the expense of Asϕ compounds. As ϕ_2 was clearly shown to be an intermediate stage, as is seen in Fig. 1. Mechanistically, the most important result of this work was the recognition that radical reactions must be involved. It was speculated that many of the arsenic atoms may form Asϕ or Asϕ_2 radicals through hot processes, which then could combine by thermal reactions with phenyl radicals produced by the recoil or by radiation damage.

An isotope effect seen in Sbϕ_3 by Maddock and Sutin[56)] was studied by Hall and Sutin[57)], whose results are shown in Table 3. Again, phenyl radicals were cited as the likely means of reforming the bonds. It was pointed out that owing to the occurrence of isomeric transitions in both of these antimony isotopes, differences in the conversion coefficients could lead to the isotopic differences.

Siekierska, Sokolowska and Campbell[58] irradiated $AsCl_3$ in benzene solution in order to test the efficacy of billiard-ball mechanisms in

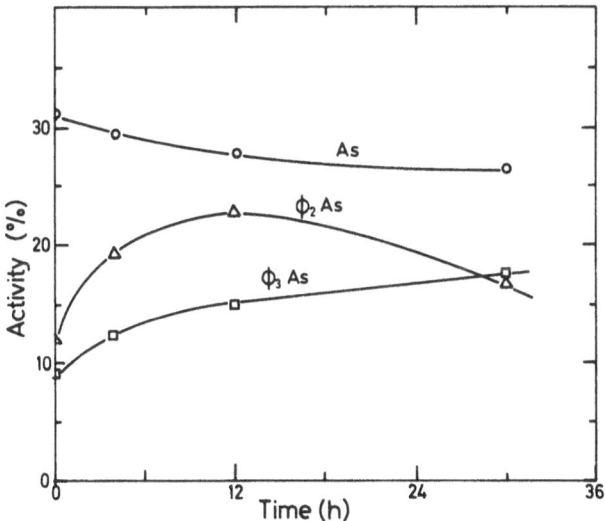

Fig. 1. Annealing curves (45 °C) for several products in neutron irradiated triphenylarsine. Note particularly the initial rise and subsequent fall of diphenylarsine derivatives (Redrawn from Maddock and Sutin[56])

Table 3. Isotope effect observed[57] in $Sb\phi_3\,(n, \gamma)$

Product	^{122}Sb	$^{122}Sb/^{124}Sb$
$Sb\phi_3$	4.4	1.37
$Sb\phi_2$	13.9	1.01
$Sb\phi$	32.7	1.06
Sb	29.7	0.88

reforming the parent molecule. While very little $As\phi_3$ was formed, the yields of $As\phi_2$ and $As\phi$ compounds totalled 20%, showing that radical reactions in benzene are reasonably efficient. Because of the number of other possible contributing reactions, it was not possible to draw definite conclusions about the effectiveness of billiard-ball collisions, other than that they seem not to be necessary in order to give carbon-arsenic bonds in liquid media. Irradiation of $As\phi_3$ in benzene solution in the presence and absence of scavengers has led[48] to a much clearer picture of what is actually going on. It was noted that in such a system, especially at low concentrations, hot exchange reactions cannot occur:

$$\text{*As} + \text{As}\phi_3 \quad \xrightarrow{\quad} \quad \text{*As}\phi_3 + \text{As}. \tag{1}$$

The results showed that the presence of oxygen caused a decrease in the yield of $\text{As}\phi_3$ and $-\text{As}\phi_2$, and an increase (not equal to the sum of the other decreases) in $>\text{As}\phi$. The decrease in clearly due to thermal scavenging effects, while the increase in $\text{As}\phi$ was ascribed to scavenging reactions such as

$$>\text{As}\phi + O_2 + H_2O \quad \longrightarrow \quad \phi\text{AsO(OH)}_2 \tag{2}$$

which would preserve a ϕAs species otherwise susceptible to further decomposition. It was observed that at higher neutron and gamma flux conditions, oxygen was no longer effective as a scavenger, owing, likely, to the probable presence of larger concentrations of phenyl radicals which could take part in hot or epithermal reactions.

These results led to the conclusion[59] that the observed products are formed through a series of consecutive reactions, occuring at different stages in the slowing down of the recoil atom: the first $\text{As}\phi$ bond is formed in the high-energy region, the second during the cooling, and the third after the cooling. The division of reaction probabilities was given as in Table 4, as representing approximate upper limits.

Table 4. Contribution from processes at various stages in the triphenylarsine benzene system[59]

	Percent ^{76}As as		
	$\text{As}\phi_3$	$-\text{As}\phi_2$	$=\text{As}\phi$
Failure of bond rupture	1—2	2	0
Hot reactions	1	5	22[a]
Thermal reactions	14	16	0

[a] Some 6% of this is susceptible to further thermal reaction with phenyl groups unless "scavenged" by oxygen to form $\phi\text{AsO(OH)}_2$.

The study of the radiochemical reactions of arsenic atoms in benzene solution was carried further[32] by comparing the product spectra of neutron irradiated AsCl_3 solutions and $^{77}\text{GeCl}_4$ solutions which have undergone beta decay. The product spectra were found to be remarkably similar, especially when considered only as to the number of $\text{As}-\phi$

bonds reformed. Only the $As\phi$ yield from ^{77}Ge seemed to be sensitive to the application of an external ^{60}Co-γ-field (up to 3.5 Mrad hr^{-1}) changing from 9% to 18% for a total dose of 25 Mrad. This result, different from that found for $As\phi_3$ solutions, must reflect a slow reaction of some inorganic arsenic compound with radiation-produced phenyl radicals. It is difficult to account with any certainty for the similarity of results from these two superficially quite dissimilar nuclear reactions. It may well be that the chemical environment is shaped more by the ionization caused by Auger cascades, which could quite accidentally be similar for ^{76}As and ^{77}As, than by other influences. Moreover, as the authors point out[32] the β ray energy is quite high (E_{max}: 42% 2.2 MeV, 35% 1.38%) so that one would expect considerable recoil from both (n, γ) and β decay processes. Perhaps more likely is that the processes may in both cases be primarily thermal reactions, which would depend little on the initial reaction.

A series of papers by Merz and Riedel describe work designed to compare radiochemical behaviour following n, γ; n, p; E.C. and β decay. Gallium isotopes are produced in most of the cases studied, but isotopes of Sn, Pb, Ge and Sb were also involved. Unfortunately, the various chromatography fractions were not well identified, so that it is not easy to draw definite conclusions from this work. Nevertheless, several things do appear to be clear. Some interesting data[60], [63] are presented in Table 5, comparing the effects of electron capture, neutron capture, and the (n, p) reaction.

Table 5

Reaction	Fraction[a]			
	Benzene	Ethanol	Water	7M HCl
^{68}Ge E.C. ^{68}Ga (solid Geϕ_4)	$\sim 2.5\%$	$<1\%$	$<1\%$	$\sim 80\%$
Ga (n, γ) ^{72}Ga (solid Gaϕ_3)	$\ll 1$	40	2	44
Ge (n, p) ^{72}Ga (solid Geϕ_4)	3.5	16.8	2.5	41

[a] The fractions were claimed to contain: benzene, Gaϕ_3; ethanol and water, more polar organogallium compounds (ethers and others); 7M HCl, atomic and ionic gallium.

The results were found not to be very sensitive to the presence of scavengers, such as atmospheric oxygen, iodine and FeCl$_3$, although the effect was sufficient to indicate clearly the involvement of radical reactions at least in the later stages. It is interesting to note here the very high ionic yield following electron capture. The authors suggest[60] that this may result from the direct formation, by electron capture and by the Auger effect, of stable Ga^{+3}.

Riedel and Merz take the 2.5 percent retention following electron capture as representing failure of bond rupture, since this decay produced no gamma ray and thus no recoil is possible.

Like Halpern, Siekierska and Siuda[77] with $^{77}GeCl_4$ in benzene, Riedel and Merz[64] found essentially the same distribution of radioactivity following β decay of $^{77}Ge\phi_4$ as by nuclear reactions, except for a uniformly higher yield of $^{77}As\phi_3$. They analyse their results for this reaction as 14% failure of bond rupture, 5% radical recombination and, in benzene solution, 4% additional reaction with radiation produced radicals.

A detailed study of the radiochemical reactions of phenylarsenic compounds has been published[70, 71, 73, 74] by Grossmann. Once again unable to effect isolation of all compounds, he was able, however, to get evidently reliable values for the sums of all compounds with one, two and three phenyl-arsenic bonds, respectively, as well as ionic arsenic and a further organic-soluble fraction which appeared to be a group of polymeric phenylarsenic compounds. Selected data from this work are given in Table 6.

Table 6. Yields of various fractions in phenylarsenic compounds[70, 71, 73, 74]. Irradiation in vacuo at $-50\,°C$, 10—20 h, thermal column (VK 3), neutron flux 4×10^9

Target	ϕ_5As	ϕ_4As	ϕ_3As	ϕ_2As	ϕAs
$\phi AsO(OH)_2$	–	–	–	0[a]	16.1[b]
$(\phi AsO)_n$	–	–	–	1[a]	24.5[a]
$(\phi As)_6$	–	–	0.5[a]	1[a]	23.8[a]
ϕAsH_2	–	–	–	1	18.8
$\phi_2AsO(OH)$	–	–	–	3[a]	15
$(\phi_2As)_2O$	–	–	0.1[a]	6[a]	19[a]
$(\phi_2As)_2$	–	–	0.3[a]	5[a]	17[a]
ϕ_2AsH	–	–	–	1[a]	17[a]
$\phi_3As(OH)_2$	–	–	2	2	13
ϕ_3AsO	–	–	1.5[a]	3[a]	16.5[b]
ϕ_3As	–	–	2.5[a]	2	16
ϕ_4AsHCO_3	–	0.7	1.0[a]	2[a]	14.6
ϕ_5As	0.2	0.2	5.2[a]	17[b]	18.9[b]

[a] Yield increases on heating.
[b] Yield decreases on heating (the heating times and temperatures varied).

Studying in addition the effect of irradiation temperature, presence or absence of air, gamma irradiation, and thermal annealing, Grossmann found results which can be summarized as follows:

> The reformation of ϕ As bonds is stepwise, much as was described by Halpern et al.[77]
>
> Both phenyl radicals and arsenic-containing radicals seem to be involved. Both can be scavenged by oxygen present either in the atmosphere or in the compound itself.
>
> On thermal annealing, ϕ As usually produces ϕ_2 As or ϕ_3 As, but sometimes gives ionic arsenic, depending on how much absorbed or constituent oxygen is present in the target. The ϕ groups may come from neighbouring molecules through exchange-like processes.

A recent study of neutron-irradiated Germanium tetraethyl was reported by Nowak and Akerman[76]. The most important result of this work was the observation of a large number of different products, separated and identified by gas chromatography. These products involve methyl and vinyl groups bonded to Germanium, as well as hydrogen and ethyl. The yields were mostly fairly low, and varied somewhat with the gamma dose received. A list of some of the compounds found, their yields, and an indication of the radiation effect are given in Table 7.

Table 7. ^{77}Ge containing products found in neutron irradiated Germanium tetraethyl (from Nowak and Akerman[76])

Compound[a]	Yield	
	At 2.5 Mrad.	At 50 Mrad.
Et_4Ge	8.3	1.0
Et_3GeMe	Not found	
Et_3GeBu	2.6	2.8
Et_3GeVi	2.2	1.6
$Et_2GeViMe$	7.4	1.2
Et_2GeH_2	3.5	3.6
$EtGeMe_2Vi$	2.0	0.3
$Et_3Ge\text{-}GeEt_3$	3.2	3.7
$(Et_3Ge)_2O$	2.9	0.7
Et_3GeOEt	2.2	2.3
$Et_2MeGe\text{-}GeMeEt_2$	1.8	1.6

[a] $Et = -C_2H_5$, $Me = -CH_3$, $Bu = n\text{-}C_4H_9$,
$Vi = -CH{=}CH_2$.

There is no indication as to whether these compounds are formed by hot or thermal reactions. Many of the products (*e.g.* the vinyl compounds and the polymers) are explainable as resulting from free radical reactions. The virtual disappearance of the parent compound at high radiation doses is attributable to the interception of the stepwise reformation by competing radical reactions. The decrease in vinyl compounds is explained[76] as being due to increased polymerisation.

To summarize the foregoing brief report on alkyl and aryl metal compounds, we can cite the following points.

1. The reactions seem to be primarily radical combinations, with some abstractions being evident. There is some question about failure of bond rupture. On the other hand it is clear that bond reformation does occur at each of the several possible stages: by hot processes, by fast, but scavenger-sensitive processes, and during thermal treatment.

2. The reformation, or formation, of molecules appears to be a stepwise process—that is, it occurs by sequential reactions which may by interrupted at certain stages:

$$As \xrightarrow{\phi} As\phi \xrightarrow{\phi} As\phi_2 \xrightarrow{\phi} As\phi_3 \tag{3}$$
$$\xrightarrow{\lfloor O} \phi AsO$$

It is interesting to note that scavenging of radicals can be done by oxygen atoms from the target compound itself[71].

A single-step displacement reaction such as

$$*As + \phi_3 As \longrightarrow \phi_3 *As + As \tag{4}$$

is not envisioned here, as it is for annealing reactions of certain cobalt(III) compounds[78].

3. The radicals are produced by various processes, although little is known for certain. A major question exists as to whether the radicals are produced by the initial bond rupture[60],[61], by recoil impact[16], by Auger ionization[79], by gamma radiation damage[76] or by normal exchange[80].

4. It is clear from the work of Siekierska, Sokolowska and Campbell[58] that the initial ligands are not essential if the solvent molecules suffice. On the other hand it is shown by Merz and Riedel[60] that solvent participation is unnecessary, and that target's own ligands are also involved.

5. It is shown by Grossmann that for his phenyl compounds the results are qualitatively the same in the presence and in the absence of strong radiation doses, yet the results of Nowak and various others show that for alkyl metals radiation effects are more important.

2. π-Ring Metal Compounds

Ferrocene was the earliest π-bonded metalorganic compound whose Szilard-Chalmers reactions were studied. Sutin and Dodson[81] did a very convincing study which showed the retention to be 12%. Thermal treatment was found to increase this to about 21%. The effect of room temperature annealing was also evident: a sample irradiated at $-196°C$ rose from R = 8.9% to R = 14.3% on standing at room temperature for 6 days. The "inorganic" yield was quite sensitive to the sample treatment —much more so than the retention as ferrocene itself. This indicated the occurrence of an additional fraction not included in the fraction studied and which evidently represents a chemically intermediate stage. This distribution of activity was also dependent on the fast neutron and the gamma ray doses. The data were interpreted as resulting from radical reactions involving $\cdot C_5H_5$ and $\cdot Fe(C_5H_5)$ radicals formed by the recoil or by gamma radiolysis.

By comparison with ferrocene, ferricinium picrate was found[82] to have a low retention. The low value of the retention (2.5%)—which includes the radioactive ferrocene—was interpreted as resulting from the characteristics of the slowing-down process in an ionic crystal, and compared with that in ferrocene, a molecular crystal. More likely, low retention should be interpreted in the light of the internal scavenging in oxygen-containing compounds, as was suggested much later by Grossmann[71].

A significant step was made[36] by neutron irradiation of ruthenocene. It was found that 20—25% of the ruthenium activity was recoverable as ruthenocene, and that also considerable rhodium activity was sublimed along with the ruthenocene. The rhodium was identified as being rhodium dicyclopentadienide, produced in high yield by the β decay of ruthenocene.

Retention in dibenzenechromium was studied by Baumgärtner, Zahn and Seeholzer[83] who found 11.8% retention in samples processed by dissolution and then sublimation, and 19.4% in samples sublimed directly. The difference was interpreted as resulting from a rapid thermal annealing during the sublimation.

Synthesis of ruthenocene from fission-product ruthenium isotopes was done[38] by neutron irradiation of U_3O_8 and $FeCp_2$ powder mixtures. It was shown that most of the ruthenocene found was actually produced by the decay of a precursor. Subsequent knowledge makes it apparent[84] that the fission product recoils formed a rhodium dicyclopentadienide whose structure was preserved through the β decay[36]. The total yield of ruthenocene reached a value of 60% under some experimental conditions and was rarely less than 40%.

A similar synthesis of ruthenocene was done by neutron irradiation of $FeCp_2$ and RuO_2 powders together. The yield of $RuCp_2$ was, however, extremely low $(0.01\%)^{37),38),40)}$.

Isotope effects* in neutron-activated ruthenocene were studied by Harbottle and Zahn[85]. Crystalline samples showed a small ($\sim 10\%$) effect, but a much larger isotope difference was found in benzene solutions. Frozen solutions were found to be similar to the crystalline samples.

The results are shown in Table 8.

Table 8

Target	Retention (%)			Isotope effect
	^{97}Ru	^{103}Ru	^{105}Ru	$^{97}Ru:^{103}Ru$
Crystals	9.5	10.7	9.9	11
2% in benzene	0.61	1.04	0.98	41
2% in benzene (frozen)	5.5	6.4	6.3	14

It was speculated[85] that perhaps the largest isotope effects may always be associated with smallest retentions. It is also of interest to note that the retention was not affected by annealing at 140 °C. However, an unidentified, sublimable product was produced by thermal treatment in amounts of up to 10% of the radioruthenium, that is, in amounts equal to the retention.

Further study of the formation of ruthenium[53] involved catching ^{103}Ru fission fragments in cyclopentadiene solutions. When monomeric C_5H_6 was used, the formation of $^{103}RuCp_2$ was very high ($\sim 90\%$), while when the dimer $(C_5H_6)_2$ was used, the yield of $^{103}RuCp_2$ was only 2.5%. This led to the suggestion that the reaction is indeed a thermal reaction having little to do with recoil or hot-atom effects. Support for this speculation was obtained by allowing monomeric C_5H_6 to stand in contact with carrier-free ^{103}Ru adsorbed (as chloride) on a glass surface. Of the radioruthenium which went into the liquid phase, up to 50% was present on subsequent analysis as $RuCp_2$, although the total conversion was quite small.

Cobaltocene and nickelocene were studied by Wheeler and McClin[86]. Retention of ^{65}Ni in nickelocene was about 65%, and increased to 92% on annealing at 100 °C for one hour. Nickelocinium ion was formed to

---

* The isotope effect is expressed as $100(R_1 - R_2)/R_1$ %, where R_1 and R_2 are the retentions of two different isotopes.

the extent of 1 %. Cobaltocene showed retention of 25 %, which increased to 40 % on annealing at 100 °C for one hour. Cobaltocene gave rise to a large fraction of active cobalt which remained in $CHCl_3$ solution, and appeared to be a "polymeric" material. Annealing of the sample caused some of this material to form $CoCp_2$. By short irradiation times, such that essentially only ^{60m}Co was formed, the retention was higher: 55 %.

Further work on nickelocene and cobaltocene was done by Ross[87], who synthesized the respective compounds using ^{56}Ni, ^{57}Ni and ^{58}Co, which decay be E.C., β^+ and a fully converted isomeric transition, respectively, all producing radioactive cobalt isotopes. The results showed retentions, after sublimation, of 84 %, 83 % and 80 %, respectively. The composition of the unsublimable residue was largely $CoCp_2^+$, except for the highly converted ^{58m}Co, where only 30 % $CoCp_2$ could be detected. This was interpreted as showing that by internal conversion the molecules are totally destroyed, by the same sort of argument as was used by Riedel and Merz[64].

The dichlorides of hafnocene and zirconocene[88] were found by Hillman, Weiss and Hahne[89] to have retentions of 15—60 % and to show very strong isotope effects. The isotope effects were even larger in benzene solution than in the crystalline targets, reaching a value of 4.13 for the ratio of the retentions of ^{180m}Hf and ^{181}Hf.

3. Carbonyls

The first study of metal carbonyls was that of Toropova[25],[90] whose objective was isotope enrichment using $Cr(CO)_6$. After dissolving the target compound in chloroform, she found nearly 90 % of the ^{51}Cr to be extracted further into 0.1 M HCl, with isotopic enrichment factors greater than 10^4. This implies retention values of the order of 10 %.

Baumgärtner and Reichold[40] prepared carrier-free $Mo(CO)_6$ in high yield by neutron irradiation of powdered mixtures of U_3O_8 and $Cr(CO)_6$. As with their preparation of $^{103}RuCp_2$, the $Cr(CO)_6$ acted only as a catcher for fission-product molybdenum (and for its precursors niobium and zirconium). The yield of 60 % found for $^{99}Mo(CO)_6$ is higher than the fractional chain yield of ^{99}Mo in fission, so that the reaction must be partly thermal, starting with molecular fragments which survive β^- decay.

An interesting study of the recoil behaviour of different nuclides in metal carbonyls was made by Harbottle and Zahn[91]. They studied $Cr(CO)_6$ and $Mo(CO)_6$ irradiated in various ways so as to produce nuclear reactions in oxygen and carbon, as well as both high and low energy reactions in Cr and Mo. The results can be briefly summarized as in Table 9.

D.R. Wiles and F. Baumgärtner

It is pointed out by these authors that the reaction of carbon atoms with oxygen atoms is very fast and quantitative. It appeared, then, that in all cases the final step was the addition of a carbonyl group to

Table 9. Retention values in $Cr(CO)_6$ and $Mo(CO)_6$
Note: These values have been assembled from Harbottle and Zahn[91]. The reader is urged to compare the original data for further information. All reactions involve high-energy recoil—$(p, p\,n), (p, p\,2n), (p, p\,n)$ or (γ, n)—unless otherwise noted.

Target	Isotope	R (%)[a]
$Cr(CO)_6$	^{11}C	30—40
$Mo(CO)_6$	^{11}C	30
$Cr(CO)_6$	^{15}O	31
$Cr(CO)_6$	$^{48,49}Cr$	65—85
$Cr(CO)_6$	^{71}Cr[b]	43, 54
$Mo(CO)_6$	$^{90,99}Mo$	70—75[c]

[a] The ranges of values encompass several values from different reactions.
[b] By (n, γ) reaction.
[c] By $(p, p\,n), (p, p\,2n)$ and (n, γ) reactions.

complete the molecule. The high retention and absence of isotope effect for the chromium isotopes suggests that the critical stages in the reformation are thermal reactions, rather than reactions dependent on the initial nuclear event.

Further insight into the reactions of metal carbonyls is provided by a study by Zahn, Collins and Collins[92] of the thermal annealing of $Cr(CO)_6$, $Mo(CO)_6$ and $W(CO)_6$ in the presence of carbon monoxide atmospheres at various pressures. Annealing curves of the usual sort were obtained, whose plateau values were temperature independent above about 120 °C. For 40 min annealing times, the values above 120 °C were 51%, 74% and 57% for $Cr(CO)_6$, $Mo(CO)_6$ and $W(CO)_6$ respectively. These values were found also to be sensitive to the irradiation conditions—being larger for higher concomitant gamma dose rates. The effect of CO pressure, up to 100 atm, was negligible at room temperature but greatly increased the effect of annealing at higher temperatures. A maximum retention of about 80% was observed for annealing at 120 °C for 2 h under 100 atm of CO. The authors suggest that the product molecule is reformed through a series of thermal reactions.

$$M(CO)_x + CO \longrightarrow M(CO)_{x+1} \tag{5}$$

where x may be between 3 and 5 after the cooling of the hot zone, and the CO is produced by recoil damage, gamma radiation, or by the applied atmosphere. They discount the possibility of CO exchange with nearby molecules.

Nickel carbonyl has been shown[93] to have a very high retention—98.7%—both in the pure liquid and as 10% solution in n-heptane. It was argued that this represents the results following essentially complete isotopic exchange. Since the exchange of CO with $Ni(CO)_4$ is known to occur quite rapidly by a dissociation mechanism, reformation of nickel carbonyl following the nuclear reaction would proceed rapidly by the reaction

$$^*Ni(CO)_x + Ni(CO)_4 \rightleftharpoons {}^*Ni(CO)_{x+1} + Ni(CO)_3. \tag{6}$$

For the same reason, this system will be self-scavenging. It thus seems unlikely that much can be learned about the hot or epithermal reactions in $Ni(CO)_4$.

Henrich and Wolf[94] have studied the formation of $Mo(CO)_6$ by catching ^{90}Mo and ^{93}Mo recoils in $Cr(CO)_6$. The Molybdenum isotopes were produced in several different reactions, so that the recoil energy varied over a wide range. It was found that the yields of $Mo(CO)_6$ with the two isotopes differed from each other, but varied only slightly as a function of initial recoil energy. These authors were also able to show that the isotope effect of about 8% is nearly insensitive to radiation received by the sample (and catcher) during the bombardment. They argued that there remains only one possible cause of this isotope effect, that is, differences in the de-excitation schemes of the product nuclei.

Groening and Harbottle[95] have found a similar isotopic difference in $Mo(CO)_6$ between ^{99}Mo and ^{101}Mo. The retention of the latter isotope is a few percent lower than that of the former. This work will be discussed further under III. B. Thermal Reactions.

In $IMn(CO)_5$, both ^{128}I and ^{56}Mn retentions have been studied[96],[97]. The retention of ^{128}I is quite high ($\sim 30\%$) likely as a result of an exchange reaction. The retention of ^{56}Mn, on the other hand, is 11%. $Mn_2(CO)_{10}$ was found to contain some 2.8% of the activity. The two retention values increase somewhat on heating of the samples at temperatures up to 80 °C, while the $Mn_2(CO)_{10}$ activity remains unchanged.

4. Polynuclear Carbonyls

$Mn_2(CO)_{10}$ was studied[43],[96] by means of dissolving the irradiated target in dilute solutions of iodine in petroleum ether, so as to scavenge any radicals which might otherwise tend to react further in the solution. By this means it was shown that in addition to the retention of 12% in $Mn_2(CO)_{10}$, a considerable fraction (4.5%) of the activity appeared as $IMn(CO)_5$. This was interpreted as indicating the presence of the $\cdot Mn(CO)_5$ radical in the crystal, especially since thermal treatment caused the abrupt disappearance of this fraction at 60%. The $Mn_2(CO)_{10}$ activity did not increase at the same time, so that the exchange reaction

$$* \cdot Mn(CO)_5 + Mn_2(CO)_{10} \rightleftharpoons *MnMn(CO)_{10} + \cdot Mn(CO)_5 \quad (7)$$

can be excluded.

Zahn found[98] that $Mn_2(CO)_{10}$ irradiated in dilute ($\sim 10\%$) solution showed very low retention—as low as 0.004%. This value, she contended, must represent the true failure of bond rupture. When these same solutions were irradiated as frozen solutions, the retention was 11.4%, quite similar to the value found for the solid.

Trinuclear carbonyls have been studied[99] with the anticipation that the retention would prove to be in some way inversely related to the molecular complexity. The values obtained were surprisingly high, despite careful chemical purification, as is shown in Table 10. It was suggested that the reformation mechanism must involve exchange reactions during and after the hot zone, starting with $M(CO)_4$ as "building blocks".

Table 10. Retentions in $Ru_3(CO)_{12}$, $Fe_3(CO)_{12}$ and $Fe(CO)_5$ following (n, γ) reaction (Narayan and Wiles[99])

Target	Species	R (Y) %
$Ru_3(CO)_{12}$	$Ru_3(CO)_{12}$	41
$Fe_3(CO)_{12}$	$Fe_3(CO)_{12}$	26
	$Fe(CO)_5$	17
$Fe(CO)_5$	$Fe(CO)_5$	41
	$Fe_3(CO)_{12}$	26

5. Arene Carbonyl Complexes

These compounds provide interesting subjects for study, since there arises the possibility of two quite different types of product molecule, in addition to the target compound: the carbonyl and the pure diarene. These possibilities are well illustrated in the study by Baumgärtner and Zahn[47] of benzenechromium tricarbonyl, as is shown in Table 11.

Production of $\phi HTc(CO)_3^+$ from $\phi HMo(CO)_3$ has been demonstrated (see below) although the production of the other possible products was not evident.

Table 11. Retention and yields of ^{51}Cr from benzenechromium tricarbonyl (Baumgärtner and Zahn[47])

Target	Product	Yield
	$(\phi H)_2Cr$	0.2%
$\phi HCr(CO)_3$	$\phi HCr(CO)_3$	10.0
	$Cr(CO)_6$	13.5

A series of studies of cyclopentadienylmanganese tricarbonyl and related compounds has provided interesting results. As with the chromium compound mentioned above, it was found that carbonyl-rich compounds are formed in yields comparable to the retention. In these compounds, however, the yield of bi-nuclear $Mn_2(CO)_{10}$ is not high, but mononuclear—$Mn(CO)_5$ compounds are prominent. The results are sumarized in Table 12.

It is noted in Table 12 that $CH_3Mn(CO)_5$ is not produced in $CpMn(CO)_3$, showing that the methyl group involved in $CH_3CpMn(CO)_3$ is the ring substituent, evidently pyrolysed or knocked off of a nearby molecule. A strange annealing behaviour in $CpMn(CO)_3$ was reported[100] in which the activity of the parent fraction (by thin layer chromatography) was observed to increase markedly during the first minutes of annealing, and then to decrease again. This effect was conjectured to be the result of the production of a highly active, unidentified compound, whose chromatographic behaviour was similar to that of the parent. Zahn showed[101] that this was indeed the case, and that the reported annealing effect was incorrect. Vasudev[102] had found two such compounds in the same target, one having a charge of +1 and the other presumably uncharged. These are listed in Table 12 as B and A respectively.

Radical diffusion processes were shown[103] to be involved in the case of $CH_3CpMn(CO)_3$ irradiated in benzene solution. Furthermore, with the same target compound irradiated with various concentrations of iso-octane the yields of $CH_3Mn(CO)_5$ was found[45] to increase, presumably as result of the increased availability of methyl radicals. The presence of $Fe(CO)_5$, far from increasing the yield by providing more carbonyls, caused the yield of $CH_3Mn(CO)_5$ to drop to zero, likely by radical competition reactions involving the methyls.

Table 12. Retentions and yields in organomanganese compounds (not including thermal treatment)

Target	Compound	Yield (%)	Reference
$CpMn(CO)_3$	Parent	12—20[a]	100
	Parent	11.7[a]	101
	Parent	7.0[a] ⎫	
	$HMn(CO)_5$	10—12 ⎪	45
	$CH_3Mn(CO)_5$	0.0 ⎬	
	$Mn_2(CO)_{10}$	0.4 ⎭	
	"A"	4—6 ⎫	100
	"B"	5—16 ⎬	
$CH_3CpMn(CO)_3$	Parent	8 ⎫	
	$CH_3Mn(CO)_5$	2 ⎬	45
	$HMn(CO)_5$	10[b] ⎭	
Fulvalene $Mn_2(CO)_6$	Parent	9.1 ⎫	
$((CO)_3Mn\ Cp\text{-}Cp\ Mn(CO)_3)$	$CpMn(CO)_3$	0.2 ⎪	46
	$HMn(CO)_5$	4[c] ⎬	
	$Mn_2(CO)_{10}$	0.25 ⎭	

[a] The ranges of values given appear to reflect a sensitivity to the different conditions used in the irradiations.

[b] Compound positively identified[44], quantitative data obtained by difference: Total $[RMn(CO)_5]-[CH_3Mn(CO)_5]$.

[c] Obtained as total $RMn(CO)_5$.

The data obtained[46] for hexacarbonylfulvalenedimanganese are a bit surprising, in that one would expect the yield of monomeric $CpMn(CO)_3$ to be higher. The high retention as parent and the low yield of the monomer can be interpreted as signifying that the radioactive manganese reacts with large molecular fragments or with whole molecules at energies too low to cause rupture of the ring-ring bond, or perhaps that the molecule is not, in fact, destroyed by the initial event.

B. Chemical Changes by β Decay and Electron Capture

The use of β decay for investigation of the chemical processes following nuclear reactions has two distinct advantages over the use of other nuclear reactions:

(i) The maximum energy of the recoiling atom is known from the maximum β decay energy, and

(ii) the probability of high charge being developed can generally be estimated from the known decay schemes and the extent of conversion of individual gamma transitions.

1. Mechanistic Studies

The most detailed studies of the results of beta decay in organometallic compounds concern $Pb(CH_3)_4$. The nuclide used is ^{210}Pb, which decays into ^{210}Bi by branched β emission: $81\% \ E_{max} = 15$ keV and $19\% \ E_{max} = 61$ keV. The 46.5 keV gamma transition is 14% converted.

After a preliminary study by Mortenson and Leighton[1] the thorough study by Edwards, Day and Overman[54] is notable. They analysed solutions of $^{210}Pb(CH_3)_4$ in benzene, octane and CCl_4 for non-volatile forms of ^{210}Bi. Similar analyses were made on gaseous $^{210}Pb(CH_3)_4$ at 10 mm pressure, both pure and diluted with He, Ne, Ar, Kr and Xe. In solution at concentrations over 5 mole percent, about 50% of the ^{210}Bi remained in a volatile form; on dilution to mole fraction 0.05, the retention fell to 18% and rose again to over 90% in very dilute solutions. The retention values in the gas phase were then practically a continuation of those in dilute solution—between 80% and 90% for the pure gas at 10 mm pressure. With helium as diluent, the retention reached its maximum of 97% and the values decreased slowly to about 90% with xenon.

Baulch, Duncan and Thomas[104] also studied the decay of $^{210}Pb(CH_3)_4$ in the gas phase from 10 mm to 0.7 mm pressure. They confirmed the earlier values for 10 mm and showed moreover that the retention decreases with decreasing pressure to less than 70% at 1 mm pressure. The form of the curve suggests that at even lower pressures the retention will fall only slightly below 70%.

Since the maximum recoil energy from β decay of ^{210}Pb is only 0.04 eV—just over thermal energies—it should be noted that recoil from β emission can have essentially no effect on the chemistry of the products. Rather, chemical behaviour is determined by the overall electron excitation from the change in Z and the conversion of the γ transitions. Investigations of the charge spectra of β decay products[105] led to the "four-fifths" rule according to which in a pure β decay without internal conversion an additional charge of several units is given to the atom through "shakeoff". This electronic excitation has been extensively discussed elsewhere[106].

From other work on $Pb(CH_3)_4$ it is known[107] that an Auger cascade connected with an L or M vacancy in the lead atom leads to the development of a charge of up to $+17$. This results in the total destruction of the molecule through a "Coulomb explosion." On the basis of the 4/5 rule and the 14% internal conversion, one can estimate that for $^{210}Pb(CH_3)_4$ the molecule should remain intact in at least 69% of the decays, corresponding to the transformation:

$$^{210}Pb(CH_3)_4 \longrightarrow {}^{210}Bi(CH_3)_4^+ + \beta^-. \qquad (8)$$

Baulch, Duncan and Thomas[104] also studied the thermodynamic stability of the $Bi(CH_3)_4^+$ produced through Eq. (8). They came to the conclusion that of the two reactions

$$Bi(CH_3)_4^+ \longrightarrow Bi(CH_3)_3 + CH_3^+ \qquad (9)$$

and

$$Bi(CH_3)_4^+ \longrightarrow Bi(CH_3)_3^+ + CH_3 \qquad (10)$$

the loss of a methyl cation is favoured over the loss of a methyl radical by about 110 kcal mole^{-1}. The thermodynamic properties thus favour the formation of the neutral volatile $^{210}Bi(CH_3)_3$ and give a further basis for the experimental value of just under 70%.

The experimentally observed minimum in retention in dilute solution remains unexplained. An explanation could lie in a scavenger action by an unknown impurity.

The influence of the decay scheme on the retention (through differences in the percent conversion of γ-transitions) was demonstrated[52, 108] by comparison of the β^--decay products of ^{210}Pb and ^{212}Pb in $Pb(C_6H_5)_3Cl$. The retention of ^{210}Bi in $Bi(C_6H_5)_3Cl_2$ was 17—19% and of ^{212}Bi about 50%. According to Nefedov, this "isotope effect" is directly proportional to the conversion coefficients of the two isotopes. Corresponding to the complement of the conversion coefficient, $1-\alpha$, the molecular structure should be preserved to the extent of 80% for the two isotopes. The probability of chemical reaction for change or preservation of molecular structure is the same for the two cases.

The effect of β^- decay of ^{210}Bi has been studied extensively[109]–[117]. Particularly interesting in $Bi(C_6H_5)_3X_2$ is the influence of the ligand X[50, 109, 110]:

$$(C_6H_5)_3BiX_2 \longrightarrow (C_6H_5)_3PoX_2^+ \longrightarrow (C_6H_5)_3PoX \qquad (11)$$

^{210}Bi decays by β^- emission of $E_{max} = 1.16$ meV with no γ-transition. Thus, the only influences are those of mechanical recoil from the β emission (maximum recoil energy 3.5 eV) and the ionization due to collective electron excitation (shakeoff).

For $X = Br$, Cl and F, the observed retentions are 73%, 33% and 5% respectively. Simultaneous with the decrease of retention, the $(C_6H_5)_2PoX_2$ increased. Since no γ transition occurs, this series clearly must reflect the influence of the halogen electronegativity on the cumulative electron excitation and the shakeoff. Evidently fluorine is least able to compensate for shakeoff and thus preserve the original structure[112].

Structure preservation at different temperatures was studied[118),119)] for the decay of ^{125}Sb in Sb(C$_6$H$_5$)$_5$ to ^{125}Te. Here a thermal sensitivity of the Te—C$_6$H$_5$ bond was revealed, which is presumably due to the intermediate state ^{125m}Te(C$_6$H$_5$)$_5^+$:

	−183 °C	20 °C
Te(C$_6$H$_5$)$_4$	73%	65%
Te(C$_6$H$_5$)$_3$X	22%	30%

$$^{125}Sb(C_6H_5)_5 \xrightarrow{\beta} \: ^{125m}Te(C_6H_5)_5^+ \longrightarrow$$

$$(12)$$

The decay of ^{131}Te in dibenzyl tellurium was studied by Halpern and coworkers[59)]. The pure compound yielded: benzyliodide 1.8%, methyliodide 36.6%, phenyl iodide 1.1%, iodotoluene 2.1%, other unidentified organic iodides 52.6% and inorganic iodine 2.4%. In benzene, carbon disulphide or carbontetrachloride solution, the yield of benzyl iodide rose to 11—18%, methyl iodide fell to a few percent, and the other values remained essentially unchanged. Addition of allyl iodide to the solution as radical scavenger caused a decrease in the yield of benzyl iodide to as low as 2.7%, depending on the allyl concentration. According to Halpern, bond breakage by recoil from β emission ($E_{max} = 1.36$ Mev) is unlikely. Likewise, it was assumed that the conversion of γ-transitions is negligible. Other measurements however, indicate 20% converted transitions. The very low yield of benzyl iodide is difficult to interpret on the basis of decay scheme data. In any case, an extension of the analysis to the hitherto unidentified organic iodine (or iodonium) compounds is necessary before anything definite can be concluded about preservation of the initial structure. Some 52—77% of the ^{131}I activity is contained in this unidentified fraction.

The β decay of ^{132}Te in Te(C$_6$H$_5$)$_2$ was investigated by Adloff[79)]. Although the decay of this nuclide is some 85% converted, the yield of ^{131}IC$_6$H$_5$ amounts to 32%, both in the pure target compound and in xylene solution. The presence of alkyl iodide as radical scavenger led to only a small difference in the yield. The inorganic fraction amounted to 52% of the ^{132}I. In this connection, it is interesting to recall the formation of 16% of iodobenzene in the β decay of ^{132}TeCl$_4$ in benzene solution. In various aliphatic alcohols, alkyl iodides with fewer carbon atoms—particularly CH$_3$I—are formed, in addition to the corresponding alkyl iodide. The authors explained these results as coming from self-radiolysis of the samples. The data can equally well be explained as resulting from recombination reactions of fragments arising through high primary charge. The mechanical recoil energy of ^{131}I from β emission may be ignored.

Chemical effects of nuclear decay have been studied in Germanium through the use of ^{68}Ge and ^{77}Ge. ^{68}Ge decays to ^{68}Ga with a 275 day half-life by 100% electron capture with no γ quanta emitted. ^{77}Ge is a β^- emitter which decays to ^{77}As with a 11.3 h half-life, by three β transitions having maximum energies of 710 keV (23%), 1379 keV (35%) and 2196 keV (42%). From this are calculated maximum recoil energies of 1.7 eV, 4.5 eV and 10.2 eV, respectively.

The results of β decay in ^{77}Ge(C$_6$H$_5$)$_4$ were investigated by Riedel and Merz[64]. A reliable statement of the reaction products formed could not be given, and the products were characterized only from their chromatographic behaviour. In the solid state, the yield of ^{77}As(C$_6$H$_5$)$_3$ was found to be 18.4—24.4%. The same yield in benzene solution amounted to 21.9%. An interesting comparison can be made between these results and those of Halpern et al.[77], who found that by decay of ^{77}GeCl$_4$ in benzene solution, radioactive As(C$_6$H$_5$)$_3$ and As(C$_6$H$_5$)$_2$$_2$O were formed in about 2.5% yield, independent of the γ dose. ^{68}As(C$_6$H$_5$)$_3$ is produced to the extent of 2—3% from K capture in ^{68}Ge(C$_6$H$_5$)$_4$.

Ross[87] has made an interesting study of β decay effects of two isotopes of Nickel: ^{56}Ni and ^{58}Ni in Ni(C$_6$H$_5$)$_2$. ^{56}Ni decays ($t_{1/2} = 6.1$ d) 100% by electron capture, while ^{58}Ni undergoes 47% β^+ emission ($E_{max} = 0.85$ meV) and 53% electron capture. The ensuing γ transition is unconverted, a fact which may influence radiochemical yields. The essential result is that in the decay of these two nuclides no volatile CoCp$_2$ is observed unless inactive CoCp$_2$ is present in the sealed capsule during the decay. With this carrier present the yield rose to 80—90%. Ferrocene did not perform this carrier function. Comparison with results from decay of ^{58m}CoCp$_2$, (100% converted) is noteworthy. Starting with uncharged ^{58m}CoCp$_2$, 70—90% of the ^{58g}Co is found as Co^{++}, 10—30% as CoCp$_2^+$ and none as volatile CoCp$_2$. Decay of ^{58m}CoCp$_2^+$ gives no CoCp$_2^+$ and only inorganic Co^{++}.

2. β Decay as a Method of Synthesis

β decay offers to pure chemistry a means of answering the question of the existence of unknown molecules. Failure to observe a particular compound in the systematics of organometallic chemistry can have several reasons. Most commonly it is because the preparative methods used are inappropriate. This is poor evidence for the non-existence of a compound, and shows as a rule only that the derived structure cannot be reached by the synthetic route chosen. β decay, on the other hand, starts from the point of having a neighbouring element already in the desired structure. After β decay, one seeks to find the decay product having *retained* its structure. Moreover, for the extremely small quantities of the product compound to be detected, it is necessary that the product nucleus must itself be radioactive. Corresponding to these conditions, previously unknown organometallic compounds of Pm, Tc. Rh, Po and Np have been synthesized and studied for the first time.

In the chemistry of cyclopentadienylmetal complexes the question was always present as to whether the uncharged compound $Rh(C_5H_5)_2$ could exist. The very stable compounds $Fe(C_5H_5)_2$, $Co(C_5H_5)_2^+$, $Ru(C_5H_5)_2$ and $Rh(C_5H_5)_2^+$ were known, but the uncharged $Co(C_5H_5)_2$ showed a very high sensitivity to oxidation, so that the existence of the homologous Rh compound was already questionable. The mass 105 decay series was especially suited for answering this question. The β^- active ($t_{1/2} = 4.5$ h) starting molecule $^{105}Ru(C_5H_5)_2$ could be easily prepared by neutron irradiation of natural $Ru(C_5H_5)_2$ followed by purification by sublimation. After the β-transformation, the criterion adopted for demonstrating the presence of the desired product compound was the volatility of the ^{105}Rh activity at $120\,^\circ C$. In fact, 15—20% of the ^{105}Rh present was found to be sublimable. In the β-decay of ^{105}Ru, 75% of the transitions involve internal conversion. If one now considers the effect of shake-off (1/5) it emerges that a theoretical maximum yield of 20% is to be expected (on the assumption that every conversion leads, by further charging, to molecular destruction). This theoretical argument is supported well in the experimental result. Subsequent attempts to synthesize the volatile Rh compound chemically led to the discovery[120] of $Rh(C_5H_5)(C_5H_6)$.

In a similar manner, the existence of $(C_5H_5)Tc(CO_3)$[37, 121] and of $Tc(C_5H_5)_2^+$ was demonstrated for the first time, starting from corresponding organometallic compounds of ^{99}Mo, which undergoes β^- decay to ^{99m}Tc. These compounds were also subsequently synthesized[122, 123], and the chemical properties confirmed.

$Po(C_6H_5)_3X$, $Po(C_6H_5)_2$ and $Po(C_6H_5)_4$ were also first prepared by β decay[124].

$$Bi(C_6H_5)_5 \xrightarrow{\beta} Po(C_6H_5)_5^+ \begin{cases} Po(C_6H_5)_3X \quad (95\%) \\ \\ Po(C_6H_5)_2 + Po(C_6H_5)_4 \quad (5\%) \end{cases} \tag{13}$$

The di- and tetraphenylpolonium could not be distinguished by the paper chromatographic method used, as was also found with $Te(C_6H_5)_4$ and $Te(C_6H_5)_2$, produced by decay of $^{125}Sb(C_6H_5)_5$. The Polonium 210 synthesis, starting with $Bi(C_6H_5)_3$ or $Bi(C_6H_5)Cl_2$ produced only $Po(C_6H_5)_3Cl_2$, and gave no trace of $Po(C_6H_5)_3$[51].

The existence and properties of $Np(C_5H_5)_3Cl$ were first determined by β-decay synthesis[125]. $^{238}U(C_5H_5)_3Cl$ served as the starting compound, being first converted by neutron irradiation to β^--active $^{239}U(C_5H_5)_3Cl$. Since the β-decay involves only a single, unconverted γ ray, one should expect a high yield. The yield of the volatile Np compound was found to be 90%. Among the cyclopentadienyl lanthanide compounds, $Pm(C_5H_5)_3$ was first prepared[126] by the β^--decay of $^{151}Nd(C_5H_5)_3$.

The synthesis of organometallic compounds as a consequence of nuclear fission is also noteworthy[37],[38],[40]. In this way, for example, a powdered mixture of $Fe(C_5H_5)_2$ and U_3O_8 gives good yields of ruthenocene and iodoferrocene on fission:

$$Fe(C_5H_5)_2 + U(n,f) \longrightarrow {}^{103}Ru(C_5H_5)_2 + Fe(C_5H_5)(C_5H_4{}^{131}I) \qquad (14)$$

Similarly, by neutron irradiation of a mixture of U_3O_8 and $Cr(CO)_6$, one gets the homologous compound $^{99}Mo(CO)_6$:

$$Cr(CO)_6 + U(n,f) \longrightarrow {}^{99}Mo(CO)_6 \qquad (15)$$

Very sensitive compounds, such as the easily oxidizable $Mo(C_6H_6)_2$, are formed when air is excluded.

$$Cr(C_6H_6)_2 + U(n,f) \longrightarrow {}^{99}Mo(C_6H_6)_2 \qquad (16)$$

In the case of nuclear fission in the presence of iodobenzene and α- or β-iodonaphthalene, despite the numerous possible diiodo compounds, one finds about 90% of the organically bound ^{131}I atoms in molecules with the original configuration:

$$\alpha\text{-}IC_{10}H_7 + U(n,f) \longrightarrow \alpha\text{-}{}^{131}IC_{10}H_7 \qquad (17)$$

$$\beta\text{-}IC_{10}H_7 + U(n,f) \longrightarrow \beta\text{-}{}^{131}IC_{10}H_7 \qquad (18)$$

The high yields—about 50%—which were observed in all cases, indicate the strong involvement of secondary fission products (i.e. those produced by β-decay of precursors).

A thorough consideration of mechanisms of formation of the organometallic products led to the conclusion[84] that the β-decay itself must be the cause of the molecule formation. Neither purely mechanical collisional substitution, nor thermal chemical reactions, nor radical reactions, nor radiation-induced reactions seem to be responsible for the synthesis reactions.

The molecule formation occurred equally well in the molten state. The only observed chemical effect was a dependence of the yield on the chemical stability of the catcher substance (as in the series $Fe(C_5H_5)_2$, $Ru(C_5H_5)_2$, $Os(C_5H_5)_2$). The formation of organometallic compounds following nuclear fission, according to these investigations, can only be regarded as connected with the collective electron excitation (shakeoff) or with a high ionization from the Auger effect following β decay.

We note a limitation of synthesis by β decay. Failing careful structure analysis the possibility remains that a new, previously unknown, structure type is present which, because of its properties being similar to those of another expected compound, may not initially be recognized as a new type of compound.

C. Some Aspects of Experimental Methods

The study of Szilard-Chalmers reactions places considerable demands on the separation methods used. These molecular compounds are only weakly polar, if at all, and thus are not easily separable. Moreover, many of the compounds and especially the radicals and other intermediates formed are not very stable, often requiring the absence of air, moisture, and even light. Combining this with the fact that many of the product species are present in extremely small amounts, we have a nearly impossible situation. Nonetheless a good deal of reliable work has been done.

Because of the instability of many of the compounds involved, it is necessary to determine the chemical recoveries in all cases. This requires the use of macro quantities (10 mg up to several hundred mg) of carriers and target compounds. This, in turn, makes it impractical to use the various thin-layer methods, such as paper and thin-layer chromatography and paper electrophoresis, although such methods have proved useful in identifying products and in checking the purity of fractions. The separation methods now most commonly used are column chromatography and sublimation.

Vacuum sublimation is a very popular method for purification of organometallic compounds, because it is so convenient and easy. The sublimation process is not very selective, however, so that it is seldom possible to separate cleanly more than one or perhaps two compounds from a mixture, while in many cases several compounds occur simultaneously in the irradiated targets. Moreover, annealing may be induced by the heating of the sample for sublimation, although this can be minimized by prior dissolution of the sample to release reactive atoms and radicals from the damaged region of the crystal. Distillation is perhaps more selective, but is seldom applicable since the small quantities used make it difficult to handle liquids.

Column chromatography is the most generally used method, and the use of several different solvents, or graded-concentration mixed solvents affords a considerable selectivity in many cases. The difficulty always remains, however, that an unidentified and unexpected species may accompany one of the carriers and give false results. This can be particularly misleading in cases where the various eluted fractions are not specifically identified, but are classified only by the polarity of the solvent.

In cases of quite volatile compounds, vapour-phase chromatography gives excellent specificity, although the usefulness of the method is limited because of the low vapour pressures and poor heat resistance of many organometallic compounds.

Best of all, especially in preliminary exploration, is to use a combination or a variety of methods. This was well shown by Narayan[99] who used various combinations of dissolution in scavenger solutions, chromatography, and sublimation. Agreement among the results of different combinations of such methods is fairly certain to indicate the absence of separation-induced reactions, and also indicates the absence of contamination by unidentified components.

A great problem is that of the stability of the compounds and more particularly of the incomplete molecular fragments. These species, whether radicals or molecules, are often very sensitive to the presence of oxygen or moisture and also have a strong tendency to adsorb on surfaces. At the same time, these primitive fragments often represent the very kind of information which we are most anxious to explore. In fortunate cases these species are transformed into more stable molecules whose precursors can be guessed with confidence. Thus, the $\phi AsO(OH)$ produced in $As\phi_3$ must clearly have resulted from the hydrolysis of a fragment containing at least one (and likely only one) $As\text{-}\phi$ bond. It is useful, where it is possible, to convert such species consciously to identifiable stable products by the use of scavengers. Grossmann[67] has used such a method to oxidise all the products from the phenylarsine compounds to more manageable compounds, without altering the number of phenyl groups attached to the arsenic atoms. Thus with $As\phi_3$ he separates only four instead of the fourteen or so products observed by earlier workers. Similarly, unstable $\cdot Mn(CO)_5$ is converted to $IMn(CO)_5$ when the irradiated target is dissolved in a dilute solution of iodine in petroleum ether. Here it was found[96, 97] that $HMn(CO)_5$ exchanges rapidly with $IMn(CO)_5$, so some of the anticipated selectivity of the reaction was lost.

As the systems studied become more complex, it will be preferable to use different, mutually exclusive, separations for specific products, rather than to attempt to separate all possible products with a single, although possibly complex process. There the difficulty will lie in the unstable species which under different procedures can give different products, and perhaps appear to be different initial species.

D. Identification of the Products

It is fruitless to attempt detailed study of a phenomenon whose products are not well identified. It is unfortunately frequently noted in the literature, especially in cases of column chromatography, that fractions are only identified as to the chemical operations which brought them to light. Fractions are identified, for example, only by the solvent used. Speculations as to the composition of the radioactive solutes in such solutions can seldom be really reliable, and the presence of an unexpected radioactive species is in such cases undetectable. It is also important in reading the literature to watch out for cases in which the chemical yields of the carriers have not been measured. Extensive decomposition can often occur on silica gel and alumina columns, especially when photosensitive or moisture sensitive compounds are used. For these reasons much of the information now existing in the literature must be regarded as only exploratory, awaiting the development of better analytical methods for separation, purification, identification and determination of the products —known or expected.

In a number of instances it has been shown that the irradiation temperature is important to control. Here it is clear that normal chemical reactions are proceeding in the target compound which will, of course, obliterate evidence of earlier reactions. Short irradiations may be done at low temperatures. Dry ice is perhaps the simplest refrigerant to use for this purpose. Even though the exact temperature of the sample may not be known, such a temperature seems to quench thermal reactions during the irradiation itself, so that these reactions may be studied later.

The atmosphere surrounding the samples is also occasionally important. Grossmann has claimed that the small but significant difference he found between vacuum and air irradiations must be proportional to the rate of diffusion of dissolved oxygen through his phenylarsine lattices. Collins has found a similar dependence on CO atmosphere in the irradiation and annealing of $(Cr, Mo, W)(CO)_6$.

III. Discussion of Results and Mechanisms

Any attempted discussion of mechanisms in the present situation must be primarily an analysis of plausible processes[127] rather than a clear demonstration of actual mechanisms. This is true for several reasons:

(i) We do not know the nature of the primary fragments and do not know whether a valid average initial species undergoes a variety of reactions or whether a number of distinct initial species undergo distinct reactions.

(ii) We only seldom have a complete picture of the product spectrum—for example, the "inorganic" fraction may in fact include a number of polymeric or polar compounds which are insoluble in non-polar solvents.

(iii) In only a few cases are there kinetic data on the effects of thermal and radiation treatment and in those cases still only on the overall result, not on any isolable reaction.

Thus, the reaction pathways suggested in the literature are highly speculative, and it must be conceded that we are still at the stage of trying to find out *what* is happening, rather than *how*.

In the face of this, then, we carry on with a fragmentary discussion of reaction pathways, hoping that the benefits of such a discussion may be an improved evaluation of the problems and a clearer view of future experimental approaches.

A. The Primary Reacting Species

In order to investigate chemical reaction mechanisms, it is useful to know the point at which the reaction starts. Regrettably, this is only rarely possible in Szilard-Chalmers studies, and is the object of one of the two most vigorous arguments in the field: whether the bonds are all broken and there is a successive rebuilding of the molecules, or these is only a partial breakage of bonds, and little rebuilding is necessary. To illustrate with $\phi_3 As$, two possible sources of radioactive parent molecules are

$$As\phi_3(n,\gamma) \longrightarrow {}^*As + 3\phi \longrightarrow {}^*As\phi_3 \qquad (19)$$

$$As\phi_3(n,\gamma) \longrightarrow {}^*As\phi_2 + \phi \longrightarrow {}^*As\phi_3 \qquad (20)$$

Although no general resolution of the question is as yet possible, it appears that each viewpoint may be valid for certain conditions.

When nuclear decay is a pure β-emission or by β with low energy, unconverted gamma-transition recoil energy is unimportant, and electronic excitation rises only—apart from the change of one charge unit by loss of the β-particle—by the sudden change of nuclear charge. It is observed in the gas phase that this collective excitation of the electron levels leads to further electron emission (shakeoff) and thus to higher charge in about 20% of the decays. In 80% of the decay events no additional ionization results and the probability for molecular survival is high, as long as the new molecule is thermodynamically stable.

When an excited state is converted by ejection of an atomic electron, a high positive charge can be produced through subsequent Auger electron emission. Within the period of molecular vibration this charge is spread throughout the molecule to all atoms, and a "Coulomb explosion" results. This primary phenomenon occurs, of course, not only as a result of β decay, but must be taken into account in all cases of nuclear reaction when deexcitation by inner electron conversion occurs[128].

With (n, γ) and other reactions more energetic than beta decay, it has for some years been an interesting question as to whether there is or is not intrinsic, primary, retention. There seems to be no question that in most cases many bonds are broken, but what is the likelihood that following neutron capture *all* bonds will remain intact? In order to investigate this question, one must[59] irradiate the target compound in very dilute solution, in the presence of much scavenger so as to suppress all radical reactions and all hot reactions with ligands of neighbouring molecules. The results must be, moreover, independent of choice of solvent, choice of scavenger, temperature, and radiation and, for a given element, independent of the choice of compound. In such cases, the measured retention can be interpreted as representing primary retention where, because of compensation of gamma-momenta and absence of Auger charging, the molecule in fact remains intact. Various other arguments might, however, be brought to bear on the data to cast doubt on this interpretation.

The study[98] of Zahn on Manganese compounds is perhaps the best example of this work. She found that in 4—5% solutions in benzene, tetrahydrofuran and acetone the retention of ^{56}Mn in $Mn_2(CO)_{10}$ was 0.005%. Moreover the retention of $CpMn(CO)_3$ in pyridine-scavenged benzene solution was found to be 0.0067%, in reasonably good agreement with the value for the carbonyl. This is convincing evidence that total failure of bond rupture is possible with manganese compounds to the extent of some 0.005%. Siekierska and Solokowska found[48] that $As\phi_3$ irradiated in oxygen-saturated benzene shows a retention of 2.1% while $AsCl_3$ in benzene gives a *$As\phi_3$ of 0.86%. From this they concluded that total failure of bond rupture leads to survival of $As\phi_3$ in 1—2% of neutron captures. The validity of this conclusion hangs to some extent on whether oxygen is sufficiently soluble in benzene to give a well scavenged system.

On the other side of the argument, it has several times been shown that the initial existence of bonds is not a requirement for final molecule production. The experiments of Siekierska and Sokolowska[48] were very clear in this respect. When $AsCl_3$ was irradiated in degassed benzene solution, phenylarsenic compounds were found up to 22% of the ^{76}As, even though $AsCl_3$ can itself act as a scavenger in this system. The formation of $^{76}As\phi_3$ was in fact only 0.86%.

Experiments by Harbottle and Zahn[91] and by Henrich and Wolf[94] used high energy nuclear reactions, such that the target atom recoiled with an energy of several MeV. In such cases, the recoil atom would certainly be a single atom, and would have a considerable initial positive charge. In these experiments, done on chromium and molybdenum carbonyls, yields of radioactive carbonyls were quite high: 50—80%. Experiments of a similar nature have been done with (γ, n) and $(n, 2n)$ reactions with analogous results, although because of the lower energies involved complete bond breakage is not so fully certain.

Combining these arguments with the observation that extensive Auger charging evidently leads to total and usually permanent molecular destruction, it can thus be concluded that molecule survival or reformation is observed to some extent when the initial species is a single atom with

low or high recoil energy, or

high initial charge when accompanied by high recoil energy, or

small initial charge without appreciable recoil energy.

The only species which apparently cannot lead directly to molecule formation is a highly charged atom with low kinetic energy. (Even these could undergo thermal reactions if they were not scavenged first.) It thus appears that almost anything can under appropriate conditions lead to molecular products of some sort.

A more difficult question arises in studies of molecules with two different ligands, such as the ring-metal carbonyls. Here one finds, as was discussed earlier, both of the homogeneous compounds, as well as the heterogeneous target compound. In the cases studied (not many as yet) there seems to be a distinct preference for the formation of the carbonyl over the diarenemetal. The question, then, is whether the bonds are selectively broken so that some carbonyls usually remain attached to the metal, or the bonds are all broken and the selectivity occurs in the reformation. (It must be conceded that the present authors do not agree on this point.) The one side of the argument contends that the metal-benzene bond (in $\phi HCr(CO)_3$) is weaker than the metal-carbonyl bonds, and is thus more likely to be broken, with at least one carbonyl ligand remaining. In this case, an initial fragment $Cr(CO)_x$ cannot lead to the formation of $(\phi H)_2Cr$, but only to either $\phi HCr(CO)_3$ or $Cr(CO)_6$. The difficulty is to reconcile such fine energy discrimination with the large total energy involved[127]. The other side of the argument suggests that if all bonds are broken, a kinetic preference for carbonyl occurs, then the observed results can be explained. This kinetic preference could arise from a combination of the greater capability of the carbon monoxide to form at least temporary bonds in all directions and the greater mobility of the carbon monoxide due to its compactness. The problem is that this viewpoint requires, in a large turbulent hot zone, total metal-to-ligand bond breakage combined with essentially total survival of the ligands themselves. On this last point hangs the second of the two most vigorous arguments in the field.

It is to be hoped that specifically designed experiments may, in the next few years, lead to an answer to this dilemma. Two experiments which are being contemplated are the following:

(i) If ^{51}Cr atoms are implanted in a target of $\phi HCr(CO)_3$ the hot-zone picture above requires that the distribution of activity be similar to that found by (n, γ) reactions. (The influence of thermal reactions must, of course, be considered in both cases.) A distinctly different pattern will discredit the hot-zone mechanism, although not necessarily prove the selective bond-breakage picture.

(ii) If one irradiates a pair of "mirror" compounds, such as $Cp_2MoH-W(CO)_5$ and $Cp_2WH-Mo(CO)_5$, the selective bond-breakage model requires that the distribution of activity be quite different for the two targets, while a similar distribution would (again after accounting for thermal reactions) indicate the existence of a certain degree of turbulence in the hot zone.

B. Thermal Reactions

Not many organometallic systems have been subjected to extensive thermal "annealing" studies— the most thorough work has been done on the phenylarsenic compounds. The earliest such study was that of Maddock and Sutin[56] on triphenylarsine. The results are in part given in Fig. 1, from which it was concluded that the reformation progresses in the stepwise fashion:

$$As\phi \longrightarrow As\phi_2 \longrightarrow As\phi_3 \qquad (21)$$

This same conclusion was reached[59] for this system by Siekierska, Halpern and Siuda who were, in addition, able to include hot reactions in their analysis, as was shown in Table 4.

The work of Grossmann[70), 71), 73), 74)] has been especially significant in this respect. In a long series of compounds with from one to
five phenyl groups bonded to each arsenic atom, he has found two quite
different types of annealing effect. In several compounds, notably triphenylarsine and phenylarsine oxide, annealing in vacuo gave a decrease
in the activity of non organically bound arsenic, and corresponding increases in the phenyl-bonded arsenic. In other compounds, notably
phenylarsonic acid and triphenylarsine oxide, as well as phenylarsine
oxide annealed in air, the "inorganic" arsenic increases on annealing, at
the expense of the phenyl-bonded arsenic. These two cases are illustrated
in Figs. 2 and 3, for the cases of diphenylarsonic acid annealed in air and
bisdiphenylarsine oxide, $(\phi_2 As)_2 O$, annealed in vacuo.

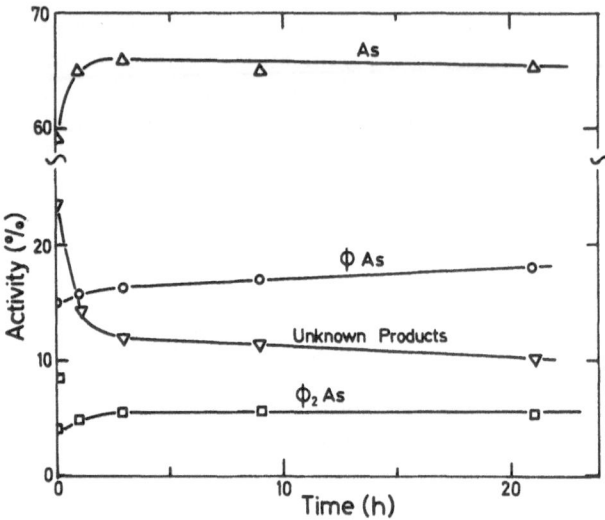

Fig. 2. Curves showing annealing effects (140 °C) on diphenylarsonic acid ($\phi_2 AsO(OH)$).
Note the rise in the yield of radioactive arsenic (Redrawn from Grossmann[71)])

Grossmann attributes these results to simple reactions of phenyl
radicals, in competition with two effects of oxygen:

(i) If oxygen is present in the target molecule (as in the
phenylarsonic acids) one may expect this oxygen to act
as a scavenger,

(ii) when atmospheric oxygen is present during annealing,
this can act as a scavenger to the extent that it is able
to perfuse through the lattice.

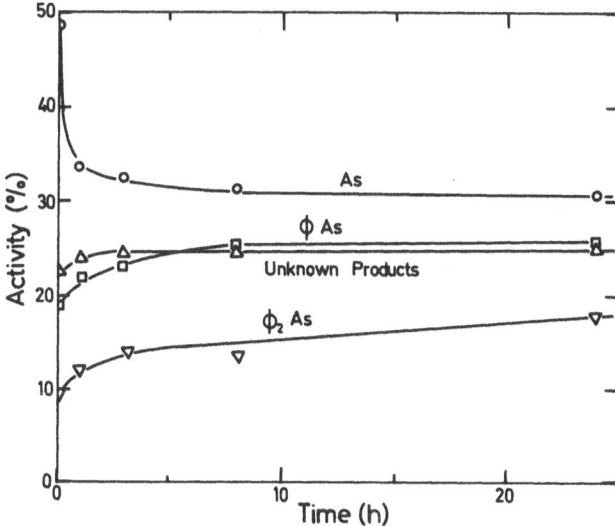

Fig. 3. Curves showing annealing effects (85 °C) in bisdiphenylarsine oxide ((φ₂As)₂O). Note the fall in the yield of arsenic (Redrawn from Grossmann[71])

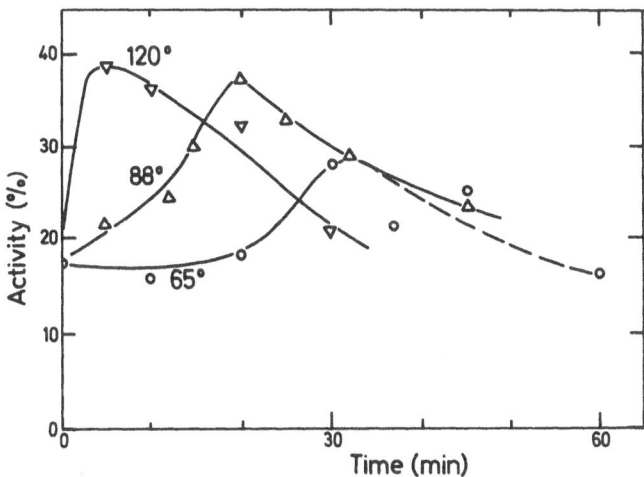

Fig. 4. Annealing data for cyclopentadienylmanganese tricarbonyl, showing an initial rise and subsequent fall in the apparent retention[100]. Other data[101] showed the retention not to have this maximum so that the extra must be an additional compound

A curious case of stepwise formation and subsequent decomposition was observed[100] on thermal annealing of $CpMn(CO)_3$, as is shown in Fig. 4. Zahn showed[101] that $CpMn(CO)_3$ undergoes only a small rapid rise to a flat plateau. Thus the effect shown in Fig. 4 must clearly result from the presence of an impurity which was carried along with the parent compound in the separation procedure. In the case cited, thin-layer chromatography in benzene was used. The compound which increases and decreases in its activity so strikingly has not yet been identified.

Thermal decomposition of an unstable species in the $Mn_2(CO)_{10}$ was shown[45] to occur rather suddenly at 55—60 °C, as is shown in Fig. 5. This species was concluded to be the $\cdot Mn(CO)_5$ radical. The

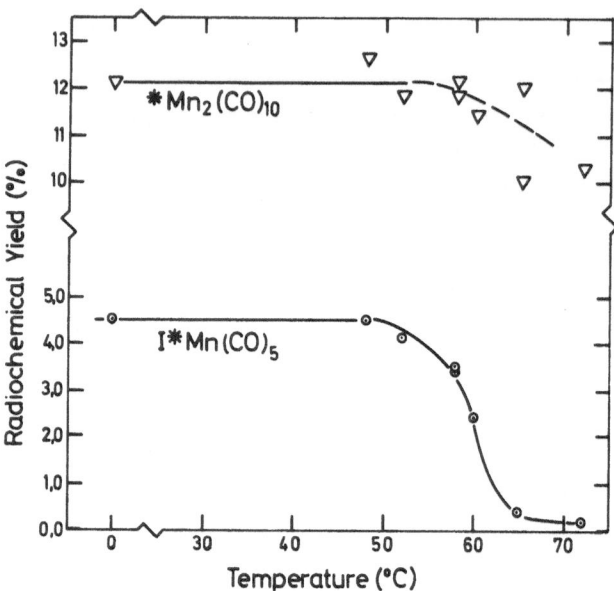

Fig. 5. Isochronal annealing data for $Mn_2(CO)_{10}$[45]. The abrupt disappearance of the $I^*Mn(CO)_5$ is attributed to the decomposition of $\cdot Mn(CO)_5$

fact that the retention value of the $Mn_2(CO)_{10}$ does not correspondingly increase at temperatures above 55 °C shows that the reaction of $\cdot Mn(CO)_5$ must be in fact a decomposition, and is not an exchange; that is, the reaction

$$*\cdot Mn(CO)_5 + Mn_2(CO)_{10} \rightleftharpoons *MnMn(CO)_{10} + \cdot Mn(CO)_5 \quad (22)$$

does not occur prominently.

More conventional annealing behaviour was found by Zahn, Collins and Collins[92] for $Cr(CO)_6$. Both in vacuo and under a 100-atm pressure of carbon monoxide, thermal treatment gave reformation of $Cr(CO)_6$ with, at least qualitatively, the usual shape of the annealing curve. This was interpreted as being a simple thermal reaction, represented as:

$$Cr(CO)_x + CO \longrightarrow Cr(CO)_{x+1} \quad (23)$$

where x is considered to be about 3 at the beginning of thermal reactions. It was concluded that the CO species involved may be either produced by radiolysis in the crystal, or provided from the carbon monoxide atmosphere by diffusion. In distinct contrast to these results for reaction of carbon monoxide with chromium carbonyl, benzene molecules do not react in the case of dibenzene chromium[83].

Information on isochronal annealing of $Mo(CO)_6$ has been given recently by Groening and Harbottle[95]. The most interesting result in this work was the clearly stepwise nature of the annealing, as is shown in Fig. 6. Curiously, not only the retention values but also the number and positions of the steps show isotropic differences. No clear explanation was offered other than the suggestion that the effect must arise from differences in the decay modes of the two excited nuclides.

Results of a recent annealing study[129] on neutron-irradiated dimeric $[CpFe(CO)_2]_2$ are given in Fig. 7. The strong effect on the $Fe(CO)_5$ is in keeping with the $Cr(CO)_6$ results mentioned above. The data for retention in $[CpFe(CO)_2]_2$ as a function of temperature are difficult to interpret unambiguously. Simple transfer of a CO molecule, such as

$$Cp_2Fe_2(CO)_2 + 2CO \longrightarrow Cp_2Fe_2(CO)_4 \quad (24)$$

ought to have the same general temperature dependence in both compounds. It is possible that this involves rather the recombination of more complex groups: even, for example, the monomeric $CpFe(CO)_2$ radical.

Activation energies of thermal reactions in organometallic compounds have unfortunately not yet been measured. It would seem that the Group VI carbonyls would offer the best possibilities for such measurements, since their reactions are not complicated by competitive reactions. Unfortunately it cannot be said with confidence that these compounds are representative of all other organometallic compounds as well.

It is generally understood[48),56),74),81)] that the reactions involved in organometallic Szilard-Chalmers Chemistry are between electrically neutral species—free radicals. This is a plausible inference in most cases,

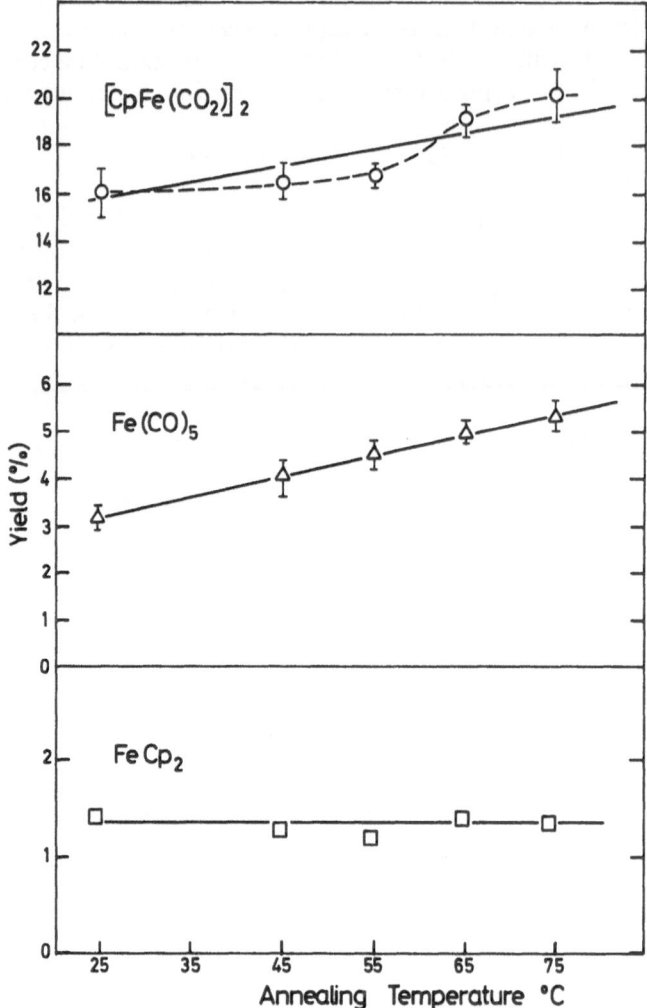

Fig. 6. Isochronal (15 min) annealing effects for various products from neutron-irradiated dicyclopentadienyldiiron tetracarbonyl ($Cp_2Fe_2(CO)_4$) (From Kanellakopulos-Drossopulos and Wiles[129)])

but it must be cautioned that the direct proof is weak or entirely lacking. The experimental evidence consists of scavenger effects in solution[98),103)] or on dissolution[43] and oxygen effects during annealing and during irradiation[48),70),71),73)−75] demonstrated reactions involving neutral species[92] radiation effects on retention[100),101),70),71),73] and radiation promotion of annealing[92]. While for thermal reactions this inference is likely safe enough, it remains possible that the species scavenged could be charged fragments or other unstable species. It would be a very

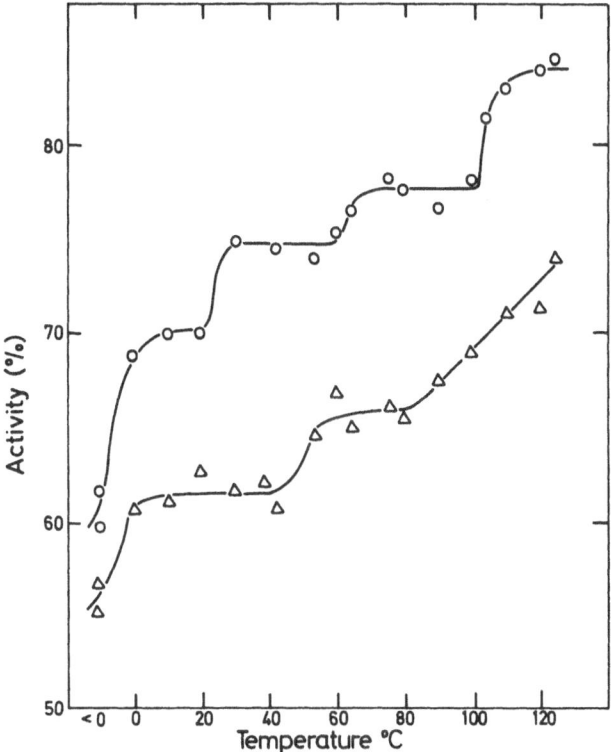

Fig. 7. Isochronal (10 min) annealing data for neutron-irradiated $Mo(CO)_6$, showing the stepwise annealing and the isotope effect (Redrawn from Groening and Harbottle[95])

useful experiment to demonstrate the actual involvement of radicals by the use of electron spin resonance, so that the radical nature of these reactions would be put on a firm basis.

One of the common tests for free radical reactions seems to be violated in these experiments, namely that they should have nearly zero activation energy. Annealing reactions in the solid state are obviously temperature dependent. However, radicals existing in solids are doubtless trapped in the lattice so that the temperature sensitive step would be radical release or radical diffusion. Studies of the diffusion of molecules in solids would be useful in this connection—for example diffusion of carbon monoxide in metal carbonyl crystals.

The origin of these radical species is also not known. It is often considered that they may result from recoil, either from the original molecule[83] or from fragments of other molecules following collision[48]. Perhaps the most commonly assumed, and the most likely, source of free radicals is radiolysis of the target compound[74],[92]. The prevalence of radiation influence on retention and on subsequent thermal treatment is convincing evidence. Perhaps the clearest case of radiation induced radical effects is that of Nowak and Akerman[76] (see Table 7), who have found a variety of reaction products in $Ge(C_2H_5)_4$ whose existence and radiation sensitivity seem only to be explainable in terms of radiolytic radical reactions and subsequent abstractions and combinations.

In this connection it seems odd that no product other than the parent compounds occurs in $(CH_3)_2Hg$ and $(C_2H_5)_2Hg$[29]. Here one would be tempted to suggest that the other expectable products have been masked through exchange reactions. This, however, seems not to be the case, since the known half times for the exchange reactions are too long to be involved here.

Dibenzene chromium was found[83] to undergo an appreciable annealing which was ascribed to the presence of a reactive benzene molecule in the near vicinity of the radioactive atom. This conclusion regarding the spatial correlation of the reacting species was reached from the γ-dose independence of the annealing (showing that bulk radiolysis is not a factor) and from the fact that samples showed no difference as to whether they were dissolved in benzene or hexane (showing that ground state molecules do not react with the incomplete radioactive fragments). It is possible to speculate that the reactive benzene molecule may be a triplet state radical somehow trapped near or partially bonded to the chromium atom.

A very interesting result on ruthenocene[53] showed that when fission product ruthenium was projected into dimeric cyclopentadiene, the yield of ruthenocene was quite low, while when monomeric cyclopentadiene was used, the yield was close to 100%. This was interpreted as involving a thermal reaction between the ruthenium atom and a cyclopentadiene monomer molecule, likely the simple displacement of an acid hydrogen.

Exchange and other similar reactions have been occasionally inferred, but seem only in a very few cases to offer a good explanation.

The high yield in $Ni(CO)_4$ is expectable in this light. Assuming a reformation intermediate $\cdot Ni(CO)_3$, we have

$$*Ni(CO)_3 + Ni(CO)_4 \longrightarrow [*Ni(CO)_3 + CO + \text{---}Ni(CO)_3]$$
$$\longrightarrow *Ni(CO)_4 + Ni(CO)_3 \quad (25)$$

which is in accord with the known mechanism for the exchange of ^{14}CO with $Ni(CO)_4$[93].

IV. Conclusions

If one is able to draw conclusions at this stage, then they must be very tentative and should best reflect present indications as to profitable next steps. The era of measuring simple retentions is past and we are at the stage of measuring trends and external influences. It is to be hoped that this stage will increasingly become a period of selecting experiments and systems in order to answer specific questions.

The next stage, which we ought to try to reach as soon as possible, is that of measurement of physical parameters of the systems so as to relate hot atom chemistry to other areas of physical chemistry. Here one would think immediately of the possible use of electron spin resonance to observe the existence and stability of the various radical species which have been postulated or inferred to occur in several systems: $\cdot Cr(CO)_x$, $\cdot Mn(CO)_5$, $\cdot FeCp$, $\cdot As\phi_2$, $\cdot \phi$ and others. One might in a similar way hope to determine the nature of the "inorganic fraction": whether it is polymerized species such as $[\text{---}Cr(CO)_x\text{---}]_n$, whether partly oxygen-scavenged species such as $CrO(CO)_x$, or whether ionic compounds such as Cr_2O_3.

Another step must be to design experiments to determine the effect of differing de-excitation modes. What are the relative effects of pure recoil, of local heating and of Auger ionization? The most promising approach, from one direction at least, seems to be the potential use of ion accelerators to bombard solid targets with radioactive metal ions. Something similar has, of course, been done through the use of high energy nuclear reactions, but the range of experimentally controllable parameters is very small. One could, for example, bombard $Cr(CO)_6$ with $^{51}Cr^+$, $FeCp_2$ with $^{59}Fe^+$, $CpMn(CO)_3$ with $^{56}Mn^+$ and so on. If the pattern of yields obtained is similar to that observed through neutron bombardment, then Auger ionization can be eliminated as a necessary mode of excitation. (The reverse experiment is not so easily conceived.) An additional item of flexibility is added in the possible use of non-isotopic projectile atoms. Here one can envisage bombarding $NiCp_2$ with $^{59}Fe^+$, $Mo(CO)_6$ with $^{51}Cr^+$ or with $^{187}W^+$, $[CpFe(CO)_2]_2$ with $^{56}Mn^+$, and the like. The roles of relative stabilities, metal ion mass, initial range and competitive reactions can all be assessed through this technique. This method, incidentally, may provide an efficient means of synthesizing a wide variety of labelled molecules of high specific activity.

The irradiation (or ion bombardment) of solid solutions, where a scavenger can be present, should also be explored further. Here it will be important to ensure that the solids are indeed solutions before conclusions can be safely drawn. It is curious to note that the yields observed in frozen solutions are in several cases very similar to the yields in the pure crystalline solutes. This suggests the possibility that the frozen targets had segregated, and that the solute was in fact present as micro crystals. (If this is the case, it may well be that a new method can be developed on this basis for making phase studies at high dilution.)

It is difficult to see beyond this last stage or in fact well into it. One can expect that the reactions will turn out to be very complex, and can only hope that they will be neither too complex to understand nor so simple as to be trivial.

Note Added in Proof. Since this review was completed, two significant experiments have been done whose results bear strongly on the question of the mechanism of molecule formation following nuclear activation.

(i) The $^{56}Fe(n,p)$ ^{56}Mn reaction has been done on $[CpFe(CO)_2]_2$ in the hope of finding $^{56}Mn(CO)_5$ and $Cp^{56}Mn(CO)_3$ produced by the high energy recoil. The yields of both compounds were very low ($< 0.1\%$)[130].

(ii) $Cr(CO)_6$ targets have been bombarded with ^{56}Mn ions at energies ranging from 4 to 20 keV, with the view to obtaining $^{56}Mn(CO)_5$. The yields have been very low ($\leqslant 0.5\%$)[131].

These two experiments, which are being extended, show quite definitely that (in these compounds, at least) pure translational "recoil" is not sufficient to cause the reactions described in this review.

V. Bibliography of General References

[1] Chemical Effects of Nuclear Transformations. I. A. E. A. Symposium, Vienna 1961.
[2] Chemical Effects of Nuclear Transformations. I. A. E. A. Symposium, Vienna 1965.
[3] Willard, J. E.: Chemical Effects of Nuclear Transformations. Ann. Rev. Nucl. Sci. *3*, 193—220 (1953).
[4] Harbottle, G., Sutin, N.: The Szilard-Chalmers Reaction in Solids. Advan. Inorg. Chem. Radiochem. *1*, 267—314 (1959).
[5] Campbell, I. G.: Chemical Effects of Nuclear Activation in Gases and Liquids. Advan. Inorg. Chem. Radiochem. *5*, 135—214 (1963).
[6] Harbottle, G.: Chemical Effects of Nuclear Transformations in Inorganic Solids. Ann. Rev. Nucl. Sci. *15*, 89—124 (1965).

7) Müller, H.: Chemische Folgen von Kernumwandlungen in Festkörpern. Angew. Chemie *79*, 128 (1967).
8) Siuda, A.: Polish Atomic Energy Commission, Bibliography to 1962. Review Report No. *6*, 1963.
9) Adloff, J. P.: Effets chimiques des transformations nucléaires. Bibliography with annual supplements. Strasbourg: Laboratoire de Chimie Nucléaire 1963.
10) Stöcklin, G.: Chemie Heisser Atome. Weinheim: Verlag Chemie 1969.

VI. References

1) Mortenson, R. A., Leighton P. A.: J. Am. Chem. Soc. *56*, 2397 (1934).
2) Herr, W.: Z. Naturforsch. *7b*, 201 (1952).
3) Melander, L.: Acta Chem. Scand. *2*, 290 (1948).
4) Schwartz, A., Rafaeloff, R., Yellin, E.: Intern. J. Appl. Radiation Isotopes *20*, 853 (1969).
5) Reichold, P., Anders, H. P.: Radiochim. Acta *5*, 44 (1966).
6) Rudenko, N. P., Schuvaeva, T. M., Merz, N. I.: Radiokhimiya *6*, 329 (1964).
7) Collins, K., Collins, C.: Nucl. Appl. *5*, 140 (1968).
8) Parker, W., Perez Alarcon, J.: Radiochem. Radioanal. Letters *3*, 223 (1970).
9) Maurer, W., Ramm, W.: Z. Physik *119*, 602 (1942).
10) Starke, K.: Naturwissenschaften *28*, 631 (1940).
11) — Naturwissenschaften *30*, 107 (1942).
12) Kienle, P., Weckermann, W., Baumgärtner, F., Zahn, U.: Naturwissenschaften *49*, 294 (1962).
13) — Weckermann, B., Baumgärtner, F., Zahn, U.: Naturwissenschaften *49*, 295 (1962).
14) — Baumgärtner, F., Weckermann, B., Zahn, U.: Radiochim. Acta *1*, 84 (1963).
15) — Wien, K., Zahn, U., Weckermann, B.: Z. Physik *176*, 226 (1963).
16) Blachot, J., Carraz, L. C.: Radiochim. Acta *11*, 45 (1969).
17) Stranks, D. R., Baker, F. B.: Inorg. Syn. *7*, 201 (1963).
18) Born, J. H.: Euratom. Report, Eur. 2209. e (1964).
19) Baumgärtner, F., in: Chem. Effects Nucl. Transformations. I.A.E.A., Vol. 2, p. 507. Vienna 1965.
20) Born, J. H.: Euratom. Report, Eur. 3282. d (1967).
21) Kahn, M.: J. Am. Chem. Soc. *73*, 479 (1951).
22) Williams, R. R., Jenks, G. H., Leslie, W. B., Richter, J. M., Larson, Q. V., in: Nat. Nucl. Energy (eds. C. D. Coryell and N. Sugarman) Ser. IV, p. 9. Radiochemical Studies: The Fission Products. Vol. 1, p. 184. McGraw-Hill 1951.
23) Spano, H., Kahn, M.: J. Am. Chem. Soc. *74*, 568 (1952).
24) Murin, A. N., Nefedov, V. D., Baranovskij, V. I., Popov, D. K.: Soviet Phys. "Doklady" (English Transl.) *1*, 719 (1956).
25) Toropova, M. A.: J. Inorg. Chem. U.S.S.R. *2*, 1201 (1957).
26) Popplewell, D. S.: J. Inorg. Nucl. Chem. *25*, 318 (1963).
27) Nowak, M.: Intern. J. Appl. Radiation Isotopes *16*, 649 (1965).
28) Colonomos, M., Parker, W.: Radiochim. Acta *12*, 163 (1969).
29) Heitz, C.: Bull. Soc. Chim. France *1967*, 2442.
30) Riedel, H. J., Merz, E.: Radiochim. Acta *6*, 144 (1966).

[31] Hoi, B., Caussé, R., Daudel, P., Flon, M., Hertzeg, C., Hoan, N., Lacassagne, A.: Compt. Rend. 228, 868 (1949).
[32] Heitz, C., Adloff, J. P.: Bull. Soc. Chim. France 1964, 2917.
[33] Wheeler, O. H., McClin, M. L.: Intern. J. Appl. Radiation Isotopes 18, 788 (1967).
[34] Baumgärtner, F., Fischer, E. O., Zahn, U.: Chem. Ber. 94, 2198 (1961).
[35] — — — Naturwissenschaften 48, 478 (1961).
[36] — — — Chem. Ber. 91, 2336 (1958).
[37] — Reichold, P., in: Chem. Effects Nucl. Transformations. I.A.E.A., Vol. 2, p. 319. Vienna 1961.
[38] — — Z. Naturforsch. 16a, 374 (1961).
[39] — Schön, A., in: Proceedings of the Conference on Methods of Preparing and Storing Marked Molecules, p. 1331. Brussels 1963.
[40] — Reichold, P.: Z. Naturforsch. 16a, 945 (1961).
[41] — Fischer, E. O., Zahn, U.: Naturwissenschaften 49, 156 (1962).
[42] — Radiochim. Acta 7, 188 (1967).
[43] de Jong, I. G., Wiles, D. R.: Chem. Commun. 1968, 519.
[44] Srinivasan, S. C., Wiles, D. R., Yang, I. C.: Inorg. Nucl. Chem. Letters 2, 399 (1966).
[45] de Jong, I. G., Srinivasan, S. C., Wiles, D. R.: Can. J. Chem. 47, 1327 (1969).
[46] — Wiles, D. R.: Can. J. Chem. 48, 1614 (1970).
[47] Baumgärtner, F., Zahn, U.: Radiochim. Acta 1, 51 (1963).
[48] Siekierska, K. E., Sokolowska, A.: J. Inorg. Nucl. Chem. 24, 13 (1962).
[49] Nefedov, V. D., Vobecky, M., Borak, J.: Radiochim. Acta 4, 104 (1965).
[50] Murin, A. N., Nefedov, V. D., Zaitsev, V. M., Gracev, S. A.: Dokl. Akad. Nauk SSSR 133, 123 (1960).
[51] — — — — in: Chem. Effects Nucl. Transformations. I.A.E.A., Vol. 2, p. 183. Vienna 1961.
[52] Nefedov, V. D., Rjukhin, Yu. A., Toropova, M. A., Melnikov, V. N., Chi Minh, L., in: Chem. Effects Nucl. Transformations. I.A.E.A., Vol. 2, p. 149. Vienna 1961.
[53] Zahn, U., Harbottle, G.: J. Inorg. Nucl. Chem. 28, 925 (1966).
[54] Edwards, R., Day, J., Overman, R.: J. Chem. Phys. 21, 1555 (1953).
[55] Maddock, A. G., Sutin, N.: Research (London) Suppl. 6, 78 (1953).
[56] — — Trans. Faraday Soc. 51, 184 (1955).
[57] Hall, R., Sutin, N.: J. Inorg. Nucl. Chem. 2, 184 (1956).
[58] Siekierska, K. E., Sokolowska, A., Campbell, I. G.: J. Inorg. Nucl. Chem. 12, 18 (1959).
[59] — Halpern, A., Siuda, A., in: Chem. Effects Nucl. Transformations. I.A.E.A., Vol. 1, p. 171. Vienna 1961.
[60] Merz, E., Riedel, H. J.: Radiochim. Acta 3, 35 (1964).
[61] — Radiochim. Acta 2, 172 (1964).
[62] Claridge, R. F., Merz, E., Riedel, H.: Nukleonik 7, 53 (1965).
[63] Merz, E., Riedel, H. J., in: Chem. Effects Nucl. Transformations. I.A.E.A., Vol. 2, p. 179. Vienna 1965.
[64] Riedel, H. J., Merz, E.: Radiochim. Acta 4, 48 (1965).
[65] Wheeler, O. H., Trabal, J. E.: J. Appl. Rad. Isotopes 21, 241 (1970).
[66] Grossmann, G., Muhl, P., Gross-Ruyken, H., Knofel, S.: Isotopenpraxis 4, 23 (1968).
[67] — Zeuner, A.: Isotopenpraxis 4, 215 (1968).
[68] — Isotopenpraxis 4, 268 (1968).
[69] — Krabbes, G., Tschernko, G.: Isotopenpraxis 4, 307 (1968).
[70] — Isotopenpraxis 5, 262 (1969).
[71] — Isotopenpraxis 5, 283 (1969).
[72] — Isotopenpraxis 5, 203 (1969).
[73] — Isotopenpraxis 5, 370 (1969).

106

[74] Grossman, G.: Radiochim. Acta *13*, 31 (1970).

[75] — Krabbes, G.: Isotopenpraxis *6*, 49 (1970).

[76] Nowak, M., Akerman, K.: Radiochim. Acta *13*, 48 (1970).

[77] Halpern, A., Siekierska, K., Siuda, A.: Radiochim. Acta *3*, 40 (1964).

[78] Nath, A., Khorana, S.: J. Chem. Phys. *46*, 2858 (1967).

[79] Adloff, M., Adloff, J. P.: Bull. Soc. Chim. France *1966*, 3304.

[80] Duncan, J. F., Thomas, F. G.: J. Inorg. Nucl. Chem. *29*, 869 (1967).

[81] Sutin, N., Dodson, R. W.: J. Inorg. Nucl. Chem. *6*, 91 (1958).

[82] Jach, J., Sutin, N.: J. Inorg. Nucl. Chem. *7*, 5 (1958).

[83] Baumgärtner, F., Zahn, U., Seeholzer, J.: Z. Naturforsch. *15 a*, 1086 (1960).

[84] — Schön, A.: Radiochim. Acta *3*, 141 (1964).

[85] Harbottle, G., Zahn, U., in: Chem. Effects Nucl. Transformations. I. A. E. A., Vol. 2, p. 133. Vienna 1965.

[86] Wheeler, O. H., McClin, M. L.: Radiochim. Acta *7*, 181 (1967).

[87] Russ, R.: Report Zfk-169, Rossendorf (1969).

[88] Wheeler, O. H., McClin, M. L.: Radiochim. Acta *8*, 179 (1967).

[89] Hillman, M., Weiss, A. J., Hahne, R.: Radiochim. Acta *12*, 200 (1969).

[90] Nefedov, V. D., Toropova, M. A.: J. Inorg. Chem. (USSR) *3*, 175 (1958).

[91] Harbottle, G., Zahn, U.: Radiochim. Acta *8*, 114 (1967).

[92] Zahn, U., Collins, C. H., Collins, K. E.: Radiochim. Acta *11*, 33 (1969).

[93] Wheeler, O. H., Trabal, J. E., Wiles, D. R.: Can. J. Chem. *48*, 3609 (1970).

[94] Henrich, E., Wolf, G. K.: K. F. K. Report No. 1067 (1969).

[95] Groening, H., Harbottle, G.: Radiochim. Acta *14*, 109 (1970).

[96] de Jong, I. G., Srinivasan, S. C., Wiles, D. R.: J. Organometal. Chem. *26*, 119 (1971).

[97] — Wiles, D. R.: Can. J. Chem. *50*, 961 (1972).

[98] Zahn, U.: Radiochim. Acta *7*, 170 (1967).

[99] Narayan, S. R., Wiles, D. R.: Can. J. Chem. *47*, 1019 (1969).

[100] Costea, T., Negoescu, I., Vasudev, P., Wiles, D. R.: Can. J. Chem. *44*, 885 (1966).

[101] Zahn, U.: Radiochim. Acta *8*, 177 (1967).

[102] Vasudev, P.: M. Sc. Thesis. Ottawa, Canada: Carleton University 1965.

[103] Yang, I. C., Wiles, D. R.: Can. J. Chem. *45*, 1357 (1967).

[104] Baulch, D., Duncan, J., Thomas, F., in: Chem. Effects Nucl. Transformations. I. A. E. A., Vol. 2, p. 169. Vienna 1961.

[105] Snell, A. H., in: $\alpha \cdot \beta \cdot \gamma$ Ray spectroscopy (ed. K. Siegbahn). Vol. 2, p. 1454. Amsterdam: North Holland Publishing Company 1965.

[106] Green, A. E. S.: Phys. Rev. *107*, 1646 (1957).

[107] Carlson, T., White, R. M.: J. Chem. Phys. *48*, 5191 (1968).

[108] Adloff, M., Adloff, J. P.: Compt. Rend. *259*, 141 (1964).

[109] Nefedov, V. D., Toropova, M. A., Gracev, S. A., Grant, Z. A.: J. Gen. Chem. (USSR) *33*, 12 (1963).

[110] — Gracev, A., Gluvka, S.: J. Gen. Chem. (USSR) *33*, 325 (1963).

[111] — Toropova, M. A., Krivochatzkaya, I. V., Kesarov, O. V.: Radiokhimiya *6*, 112 (1964).

[112] — Vobecky, M., Borak, I.: Radiokhimiya *7*, 628 (1965).

[113] — Zhuravlev, V. E., Toropova, M. A., Levchenko, A. V.: Radiokhimiya *7*, 632 (1965).

[114] — Graceva, L. M., Gracev, S.A., Petrov, L. N.: Radiokhimiya *7*, 741 (1965).

[115] — Vobecky, M., Sinotova, E. N., Borak, J.: Radiokhimiya *7*, 627 (1965).

[116] — Kirin, I. S., Graceva, L. M., Gracev, S. A.: Radiokhimiya *8*, 98 (1966).

[117] Graceva, L., Nefedov, V., Gracev, S.: Radiokhimiya *9*, 738 (1967).

[118] Nefedov, V. D., Kirin, I. S., Zaitsev, V. M.: Radiokhimiya *6*, 78 (1964).

[119] — — — Radiokhimiya *6*, 123 (1964).

[120] Fischer, E. O., Zahn, U., Baumgärtner, F.: Z. Naturforsch. *14 b*, 133 (1959).

D. R. Wiles and F. Baumgärtner

[121] Baumgärtner, F.: Kerntechnik *3*, 297 (1961).
[122] Palm, C., Fischer, E. O., Baumgärtner, F.: Tetrahedron Letters *6*, 253 (1962).
[123] — — — Naturwissenschaften *49*, 279 (1962).
[124] Nefedov, V. D., Shurawlev, V. E., Toropova, M. A., Levshenko, A. V.: Radiokhimiya *6*, 632 (1964).
[125] Baumgärtner, F., Fischer, E. O., Laubereau, P.: Naturwissenschaften *52*, 560 (1965).
[126] — — — Radiochim. Acta *7*, 188 (1967).
[127] — Zahn, U.: Z. Elektrochem. *64*, 1046 (1960).
[128] Heitz, C.: Bull. Soc. Chim. France *1967*, 2439.
[129] Kanellakopulos-Drossopulos, W., Wiles, D. R.: Can. J. Chem. *49*, 2977 (1971).
[130] Wiles, D. R.: To be published.
[131] Jenkins, G. M., Wiles, D. R.: To be published.

Received May 9, 1972

Fortschritte der chemischen Forschung
Topics in Current Chemistry

Volumes published

Springer-Verlag
Berlin Heidelberg New York
London München Paris Sydney Tokyo Wien

Organometallic Compounds

Methods of Synthesis, Physical Constants and Chemical Reactions. Second Edition. Covering the Literature from 1937 to 1964
Edited by **Michael Dub**

Volume I: Compounds of Transition Metals

Edited by **Michael Dub,** Central Research Department, Monsanto Company
XVIII, 828 pages. 1966

Volume II: Compounds of Germanium, Tin and Lead

Including Biological Activity and Commercial Application
Edited by **Richard W. Weiss,** Organic Division, Monsanto Company
XX, 697 pages. 1967

Volume III: Compounds of Arsenic, Antimony and Bismuth

Edited by **Michael Dub,** Central Research Department, Monsanto Company
XX, 925 pages. 1968

First Supplement

Covering the Literature from 1965 to 1968
Edited by **Michael Dub,** Central Research Department, Monsanto Company
XXI, 613 pages. 1972

Formula Index to Volumes I to III

Prepared by Michael Dub and Richard W. Weiss, Monsanto Company
Second edition. VII, 343 pages. 1969